Palgrave Studies in Alternative Education

Series Editors
Helen Lees
Independent Researcher
London, UK

Michael Reiss
UCL Institute of Education
London, UK

This series emerges out of a recent global rise of interest in and actual educational practices done with voice, choice, freedoms and interpersonal thoughtfulness. From subversion to introversion, including alternative settings of the state to alternative pathways of the private, the series embraces a diverse range of voices.

Common to books in the series is a vision of education already in existence and knowledge of education possible here and now. Theoretical ideas with potential to be enacted or influential in lived practice are also a part of what we offer with the books.

This series repositions what we deem as valuable educationally by accepting the power of many different forces such as silence, love, joy, despair, confusion, curiosity, failure, attachments as all potentially viable, interesting, useful elements in educational stories. Nothing is rejected if it has history or record as being of worth to people educationally, nor does this series doubt or distrust compelling ideas of difference as relevant.

We wish to allow mainstream and marginal practices to meet here without prejudice as Other but also with a view to ensuring platforms for the Other to find community and understanding with others.

The following are the primary aims of the series:

- To publish new work on education with a distinctive voice.
- To enable alternative education to find a mainstream profile.
- To publish research that draws with interdisciplinary expertise on pertinent materials for interpersonal change or adjustments of approach towards greater voice.
- To show education as without borders or boundaries placed on what is possible to think and do.

If you would like to submit a proposal or discuss a project in more detail please contact: Rebecca Wyde rebecca.wyde@palgrave.com. The series will include both monographs and edited collections and Palgrave Pivot formats.

More information about this series at
http://www.palgrave.com/gp/series/15489

Anne Watson

Care in Mathematics Education

Alternative Educational Spaces and Practices

palgrave
macmillan

Anne Watson
Department of Education
University of Oxford
Oxford, UK

Palgrave Studies in Alternative Education
ISBN 978-3-030-64113-9 ISBN 978-3-030-64114-6 (eBook)
https://doi.org/10.1007/978-3-030-64114-6

This Palgrave Macmillan imprint is published by the registered company Springer Nature Switzerland AG.
The registered company address is: Gewerbestrasse 11, 6330 Cham, Switzerland

This book is dedicated to all who love both teaching and mathematics and their integration

Preface and Acknowledgements

This book is an original exploration of what might be considered 'alternative' in mathematics education. I focus on teaching that is alternative by comparison with surrounding norms. I give many examples of teaching that share some underlying principles and similarities, manifested differently in very different contexts. Although I write with mathematics teachers, educators and advisors in England in mind, it will be of interest beyond. It is not a handbook of recipes for practice but provides background, cases and practices for consideration, all drawn from real teachers working in a range of contexts.

The early chapters of this book are based on relevant research and professional literature over three or more decades. I unfold a description of the complex context for mathematics teaching and learning in schools and elsewhere. I have tried to keep the necessary references to a minimum that might provide portals and key texts for readers. The bibliography is like the visible part of an iceberg and readers can contact me (pmtheta.com) to ask for more if they wish.

I used opportunistic snowball sampling to observe practice and interview teachers, all of whom have seen what I wrote about them and corrected details where necessary, gathering more names and contacts as I went. These provide professional evidence for the existence of teaching methods where the focus is on varieties of care for mathematics learners, justified by teachers. This method of writing bucks a trend to consider randomised control trials (RCT) as the high point of evidence in education. RCTs cannot be used meaningfully with loosely defined teaching, that is, teaching in which varied choices of action are not decided in advance but are responsive in the moment. Also, RCTs rarely focus on the successes of those whose aims and practice are different from the method under investigation. RCT results can direct us to important details about whatever is being tested, but we also learn from standing back and seeing what real teachers do and why they do it. The measures of success used by teachers in this book include progress according to external measures, such as high stakes test results and value-added scores, among more subtle educational outcomes such as the development of confident ways of thinking.

Here is an extreme, but real, example of the need for situated, qualitative evidence: if a school is wondering whether running a Forest School is a good way to spend money they would require, according to a 'what works' approach, a random selection of children to experience Forest School and a test of their social, emotional and intellectual development in a few years' time because those are the qualities that Forest School is intended to support. This is too slow a way to make urgent school resource decisions, and the measuring is hard. In this book, I value the development of professional wisdom, judgement and practice.

When I had a good draft of the book I assembled a team of teachers, advisors and educators to read the text and attend some online meetings to discuss the content, structure, and clarity and to make suggestions. While the responsibility for the content is mine, these meetings were immeasurably useful; attenders drew my attention to some omissions and over-statements and also reported on key ideas that struck them while reading. I had lost some of these in the business of writing and was happy to have them found and reinstated. Some ideas needed to be more carefully and clearly stated. Before I held these meetings I had a particular sense of purpose for the book and the meetings augmented this into the text you now have. Thanks to these colleagues, I believe the book to be of service to others. It has a collaborative element. Ways of knowing that derive from people's personal experience, including mine, permeate the book alongside academic knowledge. Peer reviewing of written artefacts is one way in which knowledge can be validated; authentic stories of practice and its effects is another.

Thanks are due to the following people who contributed to my work by either correspondence, face-to-face or online discussion. Some provided contacts and background, others worked on structure and identification of key ideas.

Alison Borthwick, Andy Tynemouth, Anushka Rao, Barbara Jaworski, Ben Sinclair, Bev Caswell, Bill Barton, Colin Foster, Florence Glanfield, Suzanne Hokin, Helen Drury, Jenni Ingram, Joan Moss, Leah Owen, Lisa Lunney Borden, Louise Price, Mary Pardoe, Matt Lewis, Matt Skoss, Michelle Tregony, Mike Ollerton, Ruth Merttens, Tom Francome.

I am most grateful to the teachers and schools whose practices I describe but who remain anonymous in Chaps. 7 and 8 and elsewhere. Although some of them did not express a wish to be anonymous I have decided that it is difficult to name some and not others, so they are all anonymous (unless I have drawn from published works) except to me … and to their students and colleagues. I felt privileged to spend time with them, in their lessons, by digital links, over coffee or a meal, so I have given them a separate index. I learn most about teaching from teachers. The teacher index also includes some named Indigenous Elders and teachers.

Profound thanks to the series editors, Michael Reiss and Helen Lees, who asked me to write something. Little did they know that I would write the book that has been in the back of my mind for 30 years but thought I could never write. Also Rebecca Wyde at Palgrave Macmillan, who shepherded me and the book into its final form.

Finally, thanks to John Mason, my husband, who took notes during some online discussions and did most of the cooking while I was writing and has been a sounding board when I thought I had gathered a new insight but needed to hear how it sounded outside my own head.

Oxford, UK
2020

Anne Watson

Contents

About the Author

Professor Anne Watson became a mathematics teacher in a comprehensive school that was socially mixed and moved to be Head of Mathematics in another comprehensive school that served one of the most disadvantaged areas in the South of England. During her school teaching she used a variety of pedagogic approaches and was an active member of the Association of Teachers of Mathematics. She then moved into teacher education and research, eventually becoming Professor of Mathematics Education at Oxford University, to which she is still affiliated in retirement. Throughout her time at Oxford she was always teaching new teachers alongside other academic work, including international work with researchers and teachers. When she retired she went to teach locally (very part-time), and since then has worked extensively with teachers, teacher educators and curriculum projects in England, Wales and beyond. Her research has focused on mathematics pedagogy, learning and curriculum—always with an underlying concern for social justice. In addition to numerous publications and edited works in research and professional journals (see pmtheta.com) she has authored and co-authored:

Watson, A., & Mason, J. (1998). *Questions and prompts for mathematical thinking*. Association of Teachers of Mathematics.

Ollerton, M., & Watson, A. (2001). *Inclusive mathematics 11–18*. Continuum.

Bills, C., Bills, L., Watson, A., & Mason, J. (2004). *Thinkers*. Association of Teachers of Mathematics.

Watson, A., & Mason, J. (2005). *Mathematics as a constructive activity: Learners generating examples*. Lawrence Erlbaum Associates.

Watson, A. (2006). *Raising achievement in secondary mathematics*. Open University Press.

Watson, A., Jones, K., & Pratt, D. (2012). *Key ideas in teaching mathematics: research-based guidance for ages 9–19*. Oxford University Press.

Abbreviations

AECG	Aboriginal Education Consultative Group
DFCS	Department for Families, Children and Schools
DfE	Department for Education
DISTAR	Direct Instruction System for Teaching Arithmetic and Reading
ECC	Every Child Counts
EHSM	European Heritage School Mathematics
GCSE	General Certificate of Secondary Education
ITS	Independent Therapeutic School
M4YC	Math for Young Children
MATs	Multiple Academy Trusts
MiCC	Math in a Cultural Context
NC	National Curriculum (England)
NCETM	National Centre for Excellence in Teaching Mathematics
Ofsted	Office for Standards in Education
PISA-OECD	Programme for International Student Assessment—Organisation for Economic Co-operation and Development
RCT	Randomised Control Trial
RME	Realistic Mathematics Education
SAT	Sistema de Aprendizaje Tutorial
STEM	Science, Technology, Engineering and Mathematics
TIMSS	Trends in International Mathematics and Science Study
TRU	Teaching for Robust Understanding

List of Figures

List of Tables

1

Introduction and Recent History

The Meaning of 'Alternative'

Alternative mathematics education—what could that mean? If I am going to write about alternatives to something, I need to be clear about the 'something'. This sets up a dichotomy between a conceptualisation of what is normal and something other, different, that can be called alternative. I have spent quite a lot of my professional life questioning and softening distinctions that can be divisive in mathematics education: traditional versus progressive pedagogy; direct teaching versus inquiry teaching; procedural versus conceptual knowledge; transmission versus construction-focused teaching; expert teacher or learner versus novice teacher or learner and so on. Softening distinctions can in itself create dichotomy: labelling differences versus celebrating similarities; identifying with a particular standpoint versus assuming diversity in a shared endeavour. This book celebrates similarities among diverse practices and assumes a shared endeavour.

A. Watson, *Care in Mathematics Education*, Palgrave Studies in Alternative Education,
https://doi.org/10.1007/978-3-030-64114-6_1

A common meaning of the term 'alternative' is education for students[1] who differ neurologically, physically or psychologically from a normal range. Instead I am using 'alternative education' to describe teaching perceived by teachers to be different from what they see taking place in the publicly funded mainstream mathematics teaching around them. This can include teachers whose practice is different from other teachers in the same school. Why would any teachers act differently to what they see as a norm? A particular kind of care that coordinates teachers' care for their students and care for mathematics is at the heart of all the examples of practice I describe in this book as 'alternative'. This is not to say that there are teachers who do not care—far from it. I take it as given that teachers of mathematics care for their students and the subject. What I found in alternative educational spaces are strong manifestations of *care for their learning of mathematics*. In doing so I am turning inside-out the title of Nel Noddings' seminal work on care: '*The challenge to care in schools: An alternative approach to education*' (1992) and instead looking at alternatives to find care in mathematics education.

I use 'mainstream' to mean: the schools that most students attend, free, usually comprehensive in intake, probably somewhere fairly near their homes, following national guidelines in terms of the mathematics curriculum, subject to national inspection, dependent on government for funding and hence concerned about achievement in national tests, even if their internal organisation of classes, teachers and timetables differ. However, this could be taken to assume homogeneity in mainstream mathematics education, which would not be a true account of practice. Homogeneity in non-mainstream educational spaces would also not be a true account.

As I write in 2020 there are some strong norms in mathematics education in mainstream publicly funded schools in England. Some of these norms are well-established over at least 100 years of mathematics teaching, such as: focusing on learning procedures by starting with examples followed by exercises. Some norms are emergent, such as: using various

[1] I use the words 'student' to refer to learners in relation to schools and teachers, 'learner' when talking more generally about learning and 'children' when talking only about learners up to about age 10.

interpretations of cognitive science as instructions for teaching. Some are imposed systemically such as: regular testing and fixed schemes of work. While not assuming that every teacher and every school teaches in the same way, it is nevertheless the case that assessment and accountability systems push schools into a fairly limited range of practices so that many students are trained primarily to pass tests rather than being educated to become competent and confident students and users of mathematics. At the time of writing, tests up to mid-secondary school are strongly focused on the performance of procedures with some attention also given to artificial short contextual questions to test application. GCSE, the final stage of mathematics for many, includes some unfamiliar questions, some multi-step questions, and also some simple modelling and data handling experiences. It would be possible to be seen as a competent teacher or a successful school by teaching in a way that focuses on doing procedures and breaks problem-solving into steps, mnemonics and stereotyped advice. Although I am talking about England here, this situation is relatively universal even when the more nuanced aims of international comparison tests are used as national guides to curriculum and practice. The usual unspoken contract between teachers and students is that if students do what teachers ask them to do, they will learn mathematics and pass their tests. By and large, teachers ask them to listen and observe something they do—some calculation, manipulation, distinguishing or reasoning—and then do it themselves. Petabytes of research publications, professional development websites and powerpoints spread advice about what to present and how to present it so that students are likely to imitate, perform and remember correctly and efficiently. The interpersonal relationships of the classroom or social medium are used to support this process.

I interpret my task to be to inquire into mathematics education practices in a range of educational spaces that do not follow these strong norms. I have sought teachers who work in the way they do, somehow differing from the norms they see and interpret around them, because their situations, mathematical cultures and principles drive them to act differently, even in small ways. In pure research, the word 'difference' should not imply deficit or disorder or 'better' or 'worse', but in my writing it is inevitable that I will be guided by my own values as well as by literature and teachers' practices.

The Shape of the Book

In this chapter, I set out the current state of mathematics education in England and the recent history that led to it. This opens a survey up to Chap. 4 of background considerations about education, cognition and the mathematics curriculum that place mathematics education within its political and sociological context, the complex nature of the subject, and the developmental and cognitive dimensions of the psychology of learning.

In Chap. 5, I examine the idea of care and develop a framework for thinking about care that embraces the considerations that have preceded it.

I find it helpful to stand back from my own context in order to better identify key parameters of situations, so in Chap. 6 I go a long way away from England to some extremely different examples of education that do not exist in UK. I do this to avoid simplistic dichotomies and instead build an understanding of possible components of alternatives to mainstream mathematics teaching. In Chap. 7, I return to England to refine a view about characteristics of alternatives to 'normal' mathematics education. Finally, in Chap. 8, I write about teachers working in mainstream publicly funded contexts whose work could be said to be 'alternative' in the light of those characteristics. My whole aim is to provide existence proofs for mathematics teaching practices that demonstrate care as central, aligned with ideas that are elaborated in Chap. 5. In Chap. 8, therefore, I hope that all teachers will find something of themselves described, and maybe something they could strive for. The teachers in that chapter employ qualities and practices that offer possibilities for everyone to some extent. As Alison Borthwick said, after reading a draft of this book: 'I am relieved that alternative does not mean militant. It can be quite subtle, small, yet powerful tweaks to otherwise mainstream pedagogy'.

In Chap. 9, I reflect and offer a synthesis of *care for the learning of mathematics* that gathers students into the subject discipline as part of who they are becoming.

Setting the Historical Scene for England

This book is written for current mathematics teaching in England, although I draw in examples from far beyond, so I shall summarise the last few decades of politicisation. This is important to do because much of the international research literature assumes a norm of so-called traditional teaching and presents it as in conflict with so-called reform teaching. This does not fit the history of mathematics education in England.

In the last three decades, mathematics in mainstream state-funded schools in England and Wales has undergone major transformations due mainly to legislation supported by centrally funded initiatives. A statutory national curriculum (NC) was introduced in 1988, with a single 16-plus mathematics qualification, GCSE, to ensure that all school students would be taught mathematics until the age of 16 according to a given conceptual progression. Students would be tested regularly with centrally designed tools, and the tests used to compare students' progress and school performance. National testing became a management tool for raising standards of mathematical achievement. Prior to this, it was possible for some students to have only limited experience of mathematics, namely clerical arithmetic, for some teachers in primary to avoid teaching much mathematics at all, and for students to be on different tracks in secondary school progressing towards differently valued qualifications.

Since then, significant public money has been spent on projects to improve learning and teaching through professional development and inspection regimes. Changes of government bring new initiatives with a ritual scrapping or reshaping of existing systems, whether or not these systems have had time to make a significant difference throughout the school.

Between 1988 and 1997, mathematics teaching and the NC were influenced by the Cockcroft report (Cockcroft 1982, para. 243) which, among other things, stated that good mathematics teaching would include:

- exposition by the teacher;
- discussion between teacher and pupils and between pupils themselves;
- appropriate practical work;

- consolidation and practice of fundamental skills and routines;
- problem solving, including the application of mathematics to everyday situations;
- investigational work.

Since some of these components would have been new to many teachers, teams of advisory teachers were set up by each local authority. There were about 300 advisory teachers and their influence varied widely—often because many schools depended on published schemes that appeared to set out 'complete' pathways for every child to follow individually (Prentice 1999). This meant that opportunities for the development of some of these features of teaching could be slim. National tests, some including portfolio assessment, were devised that recognised the value of practical, problem-solving and investigational work with the intention of creating a 'backwash' effect on teaching. After a few years, a new government initiated a return to more familiar mathematics assessments based on final written tests, using parental disquiet and administration difficulties as the reasons.

Between 1997 and 2011, the professional development landscape in schools was taken over by the National Strategy, a centrally controlled process of training and materials to support raised standards. In 1998, there was a gap of 62% between high stakes results for pupils from deprived households and those for all other pupils. According to the National Strategy summary (DfE 2011a) there had been very little direct teacher input of mathematics with many pupils working through textbook systems on their own. Schools, particularly those without any mathematics specialists, found it difficult to plan in detail from the NC alone. The Strategy essentially centralised the previous advisory network and promoted a common pedagogy that focused on whole class teaching rather than the less familiar components of teaching presented in the Cockcroft Report. The Strategy 'cascaded' common practices, particularly for primary schools, namely: timetabled daily mathematics lessons; a three-part lesson structure; direct whole-class mathematics teaching; organisational methods to support and challenge all pupils; regular practice and consolidation. It provided a detailed framework with schemes of work and lesson plans for primary schools and, later, for secondary schools. Thus, pedagogy and practice became more uniform across primary schools having the effect of limiting teaching approaches that did not have three parts to a lesson, or that involved significant amounts of groupwork and self-directed study. Because schools were subject to detailed inspection regimes, the norms of the Strategy were often interpreted by school management teams, inspectors and teachers to be the formal expectations of observed pedagogy rather than an attempt to 'level up' the least-effective teaching through more rigorous planning and interaction. Teachers who did not conform to the imposed norms sometimes had difficulty justifying their approach if there was no obvious learning within the duration of one lesson, such as when a lesson was the first part of an extended exploration.

In secondary schools before the Strategy, planning had been based typically on content coverage as set out in the NC, using textbook schemes and keeping an eye on test specifications. There were pockets and projects of interesting work in planning engaging mathematics but there was little consistency in those practices. Nearly, all schools placed students in groups or sets according to prior attainment or special learning needs and taught according to the expectations of different levels of GCSE examination, so that many, despite the NC being an entitlement for all, were excluded from some high-level topics. At its demise, the Strategy could claim that it had influenced a movement towards more explicit teaching of mathematical processes and applications, regular formative and summative assessment, the tracking of progress to identify underperformance, and more attention given to those who needed intervention to learn. The gap between attainment of those from higher and lower socio-economic groups had narrowed to 26% and an earlier gender difference had, according to national tests, been eliminated. This uniformity of teaching had achieved something in getting students' noses out of textbooks and generating more interaction with teachers, maybe at the expense of more expansive and creative pedagogies.

Throughout these two decades, there had been some desire and need to develop methods of mathematics learning beyond being taught procedures followed by individual practice and application. The year 1987 had seen the publication of 'Better Mathematics' (Ahmed 1987) which reported in the successful use of exploratory mathematics with students who had low prior attainment. Afzal Ahmed gave guidance for pedagogy that focused on developing mathematical intellect. Although this was published by government the potential for it to have national effects was overtaken by the introduction of the NC and testing, but the original NC was published with a companion volume of non-statutory guidance that offered similar ideas appropriate for teaching all students (NCC 1989).

In 2008, an independent review of mathematics teaching in early years settings and primary schools, commissioned by government from Peter Williams, was published recommending the presence of a mathematics specialist in every primary school (DFES 2008). This report also recommended continued support for Every Child Counts (ECC), a research-informed initiative that enabled children whose mathematics learning had been negatively affected by various circumstances to make better progress so that they could be educated alongside peers. Public funding for ECC was withdrawn by a new government (DfE 2011b), but echoes of its practices and effects can be found in Emily's practice in Chap. 8. The 'mathematics specialist' recommendation led to a government-funded course, focusing on a broad understanding of mathematics and its pedagogy, with 1600 teachers starting it in 2010. Funding was withdrawn shortly after, despite a positive evaluation report (DfE 2013). Continuation depends now on local funding from schools or devolved authorities; cohort size has diminished significantly.

Political ideology had stepped in, as it often does. The incoming government of 2010 judged that it was time to close the Strategy, while recognising that it had made 'a significant and positive imprint on the quality of learning and teaching in schools and settings' (DfE 2011a), and devolved resources and decision making to school level. This devolution coincided with a new ideology in educational organisation that effectively reduced the role of local authorities and allowed private–public partnerships to take over vast swathes of school groupings, not always geographically or philosophically coherent, and run their own systems that allow varied levels of teacher and school autonomy. Devolution to school level is therefore often a myth, since governance of a school chain can impose its own requirements. Of course this political move would have clashed with a central advisory structure and in theory might have opened the way to more variety of practice such as, perhaps, digitally orientated exploratory mathematics teaching, collaborative project-based approaches, and other approaches that were being developed in independent schools and several European countries.

But before I look at practice post 2010 it is useful to observe that, while all this was going on, the effect of international comparison tests on policy had increased; they have become a yardstick against which many countries evaluate the success of public policies in relation to school mathematics. It is usual for English politicians to use national achievements in such tests as reason to change educational policy, but this affects England only, because national devolution has resulted in England having a particular educational regime while other nations of the UK have their own. That is why this book is about what is possible in England while its content has broader application.

There are two main international tests to which politicians refer: Trends in International Mathematics and Science Study (TIMSS) and the Programme for International Student Assessment run by the Organization for Economic Cooperation and Development (PISA or PISA-OECD). They vary slightly in what they do, TIMSS being more geared towards providing information about school mathematics achievement at ages 8 and 14, PISA leaning more towards what is important for international economic development and focusing on 15-year-olds. Over six testing cycles England's performance in TIMSS has improved (TIMSS 2015) and made among the greatest gains of all countries during the years of the Strategy.

In the 2009, PISA tests UK 15-year-olds scored round about the average of OECD countries, with a relatively large gap between boys' and girls' performance (in contrast to the Strategy claims of having closed the gap). UK also had one of the biggest differences in achievement related to socio-economic measures despite some success in reducing this. One fifth of students were only capable of answering well-defined, familiar questions, and the proportion of highest achieving students was half the OECD average, and 6% of the highest achievers in Shanghai, despite the English practice of separating students by setting for appropriate teaching. Unsurprisingly, the incoming government in 2010 used the PISA results to disband the Strategy and rethink how mathematics education, indeed the whole curriculum, needed to be changed. The latest PISA tests

(DfE 2019) at the time of writing were performed on 15-year-olds who would have had a maximum of three years of the new curriculum and had not experienced the extensive changes that had been made in primary schools. England's rank order and scores had improved significantly and brought it into line with many other European countries that had previously been ranked higher. The gap between lower and higher scores was reduced, again bringing it into line with several other European countries. Whether this was due to the 2018 test being more in line with England's own national tests, or a national mood for raised expectations, or better teaching, or a sampling bias due to a high initial refusal rate (Jerrim 2019) is debatable. The same cohort did not show a significant increase in national test scores however. Interestingly, in national tests girls and boys performed similarly, but in PISA which is more problem-focused with fewer predictable questions a significant gender gap exists, with boys doing better.

In 2012, PISA performances in different kinds of task were compared. Relative to their overall score, students in England did well in tasks that required them to interpret, apply and evaluate; average in employing mathematical concepts, facts, procedures and reasoning, and weakly in formulating situations mathematically. 13% of the variance of scores was associated with socio-economic status. It might be expected that a government using PISA as a reason for change might also use PISA to identify strengths and weaknesses and proceed accordingly. For example, if one fifth of students in 2009 could only answer familiar procedural questions so maybe more needed to be done to develop formulation skills. However, by this time the government had set out on a different course of action that, while being concerned with the socio-economic status gap, was not going to capitalise on the relative strengths of students in England so much as replace everything with a different ideology of common access to a 'knowledge-rich' curriculum, meaning one in which specified details of knowledge are explicit and remembering these is important, the argument being that without knowledge problem-solving does not have adequate tools.

It could have abandoned the recent problem-focused version of the NC it had inherited and returned to the original version that was dense with specified mathematical knowledge and was still being used by teachers. Instead it embarked on a total rewrite, claiming common knowledge as a new idea (e.g. Hirsch 1987; Young 2014). This was part of an overt generic ideological drive to ensure that school students were given a store of basic English cultural facts throughout their education, but in mathematics little is different from the earlier NC, or the curricula of other countries, apart from some content appearing at younger ages (DfE 2012).

Michael Young, the sociologist whose book about powerful knowledge (2014) was used to support this move, later distanced himself from its implementation saying: this policy might be 'superficially easier to implement' but 'can lead on its own, to an over-emphasis on memorisation' (Young 2019). This anticipation has turned out to be correct in mathematics, for which, while memory is important, adaptive reasoning is also crucial to building conceptual understanding and solving complex problems in and out of mathematics (Kilpatrick et al. 2001).

The government had first promised to step back from centrally managing pedagogy and hand autonomy to teachers and headteachers, however significant money is spent nationally on promulgating particular approaches in mathematics. A National Centre for Excellence in Teaching Mathematics was set up in 2006, funded as a hub for professional knowledge and support services, including grants for teacher-led research, online journals for information and discussion, and recognition of quality for providers of professional development. It could have provided the basis for a devolved process of mathematics teaching development. However, at the time of writing it is totally funded by the DfE and many (but not all) of its current projects at the time of writing act as conduits for DfE policy in relation to pedagogy.

Writing in 2020, the full breadth of school mathematics is taught to all students as a major priority, more or less in parallel with each other, until upper secondary level. Mathematics results in the national tests play a major role in how schools are judged. The landscape of school mathematics is, therefore, orientated towards high-stakes test results for all while low mathematics and English test results can result in enforced changes of school governance. The key to raising performance and narrowing attainment gaps between social groups is seen by government to be raising teacher expectations of all pupils' performance by using inspection, funding and testing regimes and a mastery approach that keeps all students in line with each other while learning core ideas. To a great extent, these new norms move England further into the philosophy that all students should have the same access to core mathematical ideas—an equality of access that could be avoided during the Strategy and preceding years. During the Strategy lifetime all students were taught a broad mathematics curriculum—no longer could some students be limited to clerical arithmetic—but at very different paces, so that some would finish school without having much idea about mathematics apart from numerical processes, simple reasoning about shapes, and maybe some graphing and interpretation of data. Although the government lacks the legislative power to impose a 'same mathematics for all' approach, it uses NCETM, textbook subsidies and other means as levers to make it clear that widely differing relative achievement across social groups (racial, gender, language, economic) is no longer acceptable, and it does this not only through the testing regime but also through the kind of pedagogic advice and professional development it funds. Far from stepping back, this has the appearance of constructing a 'normal pedagogy' to which many teachers and schools think they have to conform. The norms are heavily influenced by several years of government-funded exchange projects with Shanghai and government support for mastery orientated and direct instruction approaches, where these words have various meanings. In high-performing countries where these are the established norm there is time and money and staffing given to undertake extra training and teaching, while in England there is little extra money and time. These set up contradictions, such as whether can teachers combine whole class teaching with the very wide range of capabilities with which children enter their school phase, and whether teachers can combine the desire to move into new concepts only when previous concepts are well understood with a testing regime that requires everyone to 'reach a particular level' by a particular time. In addition, a generation of teachers who had experience of including inquiry and exploratory approaches in their teaching, influenced by the Cockcroft Report, thus offering expertise in other kinds of pedagogy, has largely retired.

In September 2019, after nine years of this government's regime, the Children's Commissioner pointed out that nearly 100,000 young people,

18% of the cohort, left school without the level of qualifications that would give them access to future employment and study. After a period when this proportion was falling there had been a 28% rise over the previous four years. The DfE replied that standards were generally rising, and the rise in unqualified school leavers was due to removal of qualifications in 2013/2014 that 'did not serve children well' (Weale 2019). The reduction in variety of qualifications is part of the project to have the same expectations for all. As I write, the habit of removing failing students from the statistics by taking them off school rolls has become a national scandal; at least 2900 schools 'losing' pupils before final examinations (Ofsted 2019). A national inspection report, *Mathematics made to measure* (Ofsted 2012), paints a picture that might help to explain the fall, since many of these failing students were in school during those years. Mathematics inspectors consistently pointed to certain weaknesses in the teaching of students who were not achieving what was expected. The report is based on surveys of 320 schools and 1670 teachers, and notes that the proportion of students who are underachieving at age 7 doubled by age 11, and almost doubled again by 16. They looked at the steps that schools take to raise attainment for these students. Since most of the students in this category are also socio-economically disadvantaged it is also worth seeing what PISA-OECD says. Across the OECD countries they found that disadvantaged students tended to have more low-level skills-based teaching than others and were more likely to have been given different work to do, maybe independently from the teacher with sporadic support. Is this what has been happening in England? Ofsted pointed to a need for frequent, timely and targeted intervention before and after lessons so that children can join in with what their main class is doing and for consistent approaches across years, teachers and teaching assistants so that mathematics is coherent, such as through making links between topics and using consistent forms of representation. By implication this means that too many students were not getting this kind of experience, and these would grow older to become those who are 'lost' before the examinations for which schools are accountable.

The gradual privatisation of school governance has created groups of schools, such as multi-academy trusts (MATs), that have their own internal directives on pedagogy, sometimes generic across all subjects rather than specifically mathematical, sometimes very tightly structured, sometimes granting wide autonomy to teachers. It would be tempting to say that therefore there are no national norms, but the inspection, testing and examination systems shape the norms and schools, or groups of schools, decide how to address these. For example, is everything focused on test results and the expectations of inspectors, or is there a more holistic view of education that puts these measures of performance among other aims?

References

Ahmed, A. (1987). *Better mathematics: A curriculum development study based on the low attainers in mathematics project.* London: HM Stationery Office.

Cockcroft, W. H. (1982). *Mathematics counts.* London: HM Stationery Office.

DfE. (2011a). *The national strategies 1997–2011: A brief summary of the impact and effectiveness of the national strategies.* https://assets.publishing.service.gov.uk/government/uploads/system/uploads/attachment_data/file/175408/DFE-00032-2011.pdf. Accessed 1 July 2020.

DfE. (2011b). *Every child counts: Independent evaluation.* https://assets.publishing.service.gov.uk/government/uploads/system/uploads/attachment_data/file/182404/DFE-RR091A.pdf. Accessed 11 Apr 2020.

DfE. (2012). *Review of National Curriculum.* https://assets.publishing.service.gov.uk/government/uploads/system/uploads/attachment_data/file/184064/DFE-RR178.pdf

DfE. (2013). *Evaluation of the Mathematics Specialist Teacher programme.* https://www.nfer.ac.uk/publications/pmsz01/pmsz01.pdf

DfE. (2019). *Achievement of 15-year-olds in England: PISA 2018 results.* https://assets.publishing.service.gov.uk/government/uploads/system/uploads/attachment_data/file/855985/PISA_2018_England_national_report_accessible.pdf. Accessed 11 Apr 2020.

DFES. (2008). *Independent review of mathematics teaching in early years settings and primary schools: Final report.* http://dera.ioe.ac.uk/8365/7/Williams%20Mathematics_Redacted.pdf. Accessed 11 Apr 2020.

Hirsch, E. (1987). *Cultural literacy: What every American needs to know.* Boston: Houghton Mifflin.

Jerrim, J. (2019). *Is England's PISA 2018 data reliable?* https://ffteducationdatalab.org.uk/2019/12/is-englands-pisa-2018-data-reliable/. Accessed 11 Apr 2020.

Kilpatrick, J., Swafford, J., & Findell, B. (2001). *Adding it up: Helping children learn mathematics*. Washington, DC: National Academy Press.

NCC, National Curriculum Council. (1989). *Mathematics: Non-statutory guidance*. London: Her Majesty's Stationery Office.

Noddings, N. (1992). *The challenge to care in schools: An alternative approach to education*. New York: Teachers College Press.

Ofsted, Office for Standards in Education (Ofsted). (2012). *Mathematics: Made to measure*. https://dera.ioe.ac.uk/14522/7/Mathematics_made_to_measure%5B1%5D_Redacted.pdf. Accessed 11 Apr 2020.

Ofsted, Office for Standards in Education (Ofsted). (2019). *Off-rolling: Exploring the issue*. https://www.gov.uk/government/publications/off-rolling-exploring-the-issue. Accessed 11 Apr 2020.

Prentice, J.R.R. (1999). The curriculum for 7-11 year olds. London: Sage Publications.

TIMSS. (2015). *TIMSS 2015 encyclopedia*. http://timssandpirls.bc.edu/timss2015/encyclopedia/countries/italy/the-mathematics-curriculum-in-primary-and-lower-secondary-grades/. Accessed 11 Apr 2020.

Weale, S. (2019). *'Shameful rise': 18% of children now leave school as low achievers*. https://www.theguardian.com/education/2019/sep/20/shameful-rise-18-of-children-leave-school-without-basic-qualifications. Accessed 11 April 2020.

Young, M. (2014). What is a curriculum and what can it do? *Curriculum Journal, 25*(1), 7–13.

Young, M. (2019). *Cultural capital and curriculum*. http://ioelondonblog.wordpress.com/2019/09/09/cultural-capital-and-curriculum-will-ofsteds-new-framework-encourage-better-education-in-our-schools/. Accessed 11 Apr 2020.

2

Mathematics Schooling and the Aims of Education

Differences in Views of Education

In the rest of this book, I present the complexities of teaching and learning in educational spaces in which freedom is given to, or taken by, teachers to act in ways that do not match a perceived norm within a particular jurisdiction and can, therefore, be described as 'alternative'. There will be varied ways in which mathematics is taught in different contexts with a range of overarching educational aims and purposes, some conventional and others not so. I identify how these different ways show care for students' cognition and development in mathematics. I draw on a purposeful sample of case studies from opportunistic encounters with teachers who were identified by others, or self-identified, as teaching in ways that went beyond the norms around them either institutionally or in the wider world of mathematics teaching. The aims of the education systems and perspectives within which the teachers work vary from the free development of young people to become well-functioning, happy and fulfilled individuals to the production of adults who will contribute to the state through their economic, humanitarian and social activity (Watson 1999).

A. Watson, *Care in Mathematics Education*, Palgrave Studies in Alternative Education,
https://doi.org/10.1007/978-3-030-64114-6_2

In 1991, Paul Ernest published a seminal work *The Philosophy of Mathematics Education* in which he described how different groups of ideas about mathematics and its learning and assessment could be clustered to show influences on mathematics education and hence the nature of mathematics in educational contexts. He named each cluster after a type of ideologue who might value the nexus of views and practices: industrial trainer, technological pragmatist, old humanist, progressive educator, public educator. Each ideologue would hold compatible views about politics, the nature of mathematics, moral values, society, childhood, ability, learning, teaching, assessment and social diversity. However, it cannot be assumed that all teachers, schools or policies fit neatly into one category because teachers' enactment of beliefs is always hedged around with institutional variables and also with individual limitations. Teachers may not know how to enact their beliefs. History has overtaken Ernest's categorisation of ideologues to some extent. Currently government direction in England, while based on an understanding that elementary mathematics is a body of facts and methods that can be easily tested, also recognises systemic underachievement in certain social groups, has a flexible view of ability and sees education as a vehicle for social change. While skill acquisition is seen as paramount in childhood, problem-solving capability is an explicit aim for older school students. Many of these stances would be seen as incompatible or even contradictory in Ernest's typography, but in my view his recognition of the ingredients of *how* mathematics learning is shaped still stands and provides a raw framework for understanding different influences in mathematics teaching, for example through implicit messages to students through the actions of teachers and the ways in which schools are organised. Often these are the result of pragmatic adjustments between personal beliefs and institutional structures and instructions. They may not reflect the beliefs of teachers, nor the intended stances of institutions, but something in between which could either be a conflict or a mash-up.

The Nature and Value of Mathematical Knowledge

Mathematical tasks define what it means to do mathematics, so tasks which afford different forms of activity shape different views of the subject. To show how differences in the nature of mathematical work might show up in mathematics lessons I show two sets of questions (see Table 2.1). Each set purports to address one mathematical idea, but the questions do this in different ways with different assumptions that manifest different aspects of what it means to do mathematics. The questions might be used either to initiate learning or to assess learning; both uses can influence learners' views of mathematics.

Mathematics is not neutral in its effect on learners' lives. First, a recognised mathematics qualification, however it is constructed, is necessary for many employment and educational routes. Further than that, doing mathematics involves interpretation, such as one might need for any kind of problem-solving, and can be emancipatory for students by allowing them to use pragmatic forms of knowledge they have developed outside school, or might develop for the specific purpose of solving a particular problem (Brown 1999). For example, the questions in set A Table 2.1 might be answered by applying known formulae or by understanding what 'miles per hour' means and working answers out in an ad hoc fashion based on the realistic meanings in the questions. These questions could be used either to introduce or apply symbolic representation of the standard formula: 'distance = speed × time' for constant speed. It could be argued that using the questions to introduce the concepts could help students to develop generic modelling capabilities. The first question encourages thinking about the relationships between distance, speed and time; the second encourages thinking about when and how to apply a known method by identifying underlying structure. Some of the questions would be harder to do than others partly because of the ways in which variables appear in the text, so whether a formula is known or not there is still an issue of sorting out what to do by identifying variables, constants and relationships; this is the essence of modelling, not merely a useful exam tactic.

Table 2.1 Two sets of questions that address similar content in different ways

Set A: context and mathematics	Use for teaching/learning	Used for assessment
Mim walks at 3 miles per hour. How long will it take to walk 0.75 miles? State your answer as a fraction of an hour	There are various ways to tackle this available whether students have a formula or not; could derive a general method, but 0.75 is ¾ so it is relatively straightforward to calculate the time. The need to state time as a fraction is superfluous unless it is used to draw attention to using fractions in calculations	Interpretation of worded questions; maybe apply journey formulae and then use knowledge of time and fractions, thus testing several concepts
I walk to the shops at 3 miles per hour. The shops are 1.5 miles away. My shopping takes 5 minutes. I catch the bus back without having to wait for it. It goes at an average of 10 miles per hour. How long is my shopping trip in minutes?	The question could lead to the discussion of the value of diagrams to keep track of the journey and the need to keep a tally of duration. Deduce the calculations required or apply known formulae. What changes (speed) and what stays the same (distance)?	Tests several procedures: Journey calculations and adding time intervals. Requires imagery and modelling of the whole journey, with good reading skills
A fitness test asks me to walk at 3 miles per hour for at least 30 minutes. I know that it is 1.8 miles from my house to the railway station. Can I use this to test my fitness?	Interpret a question. Deduce how to calculate or apply known formulae. Extend the situation beyond the test. Maybe use local data; cross-curricular links to geography and PE	Ignore extra information; avoid personal knowledge; identify key values. Apply formulae to make a sound judgement. Interpret 'at least'
To improve my fitness I shall increase my walking distance by 0.1 miles per day, starting from 1.5 miles at 3 miles per hour, and keep the time to half an hour exactly. When will I reach 4 miles per hour?	Decide how to represent this: Graph and/or algebraic solution by analysing variables and formulating relationships. Discussion of most useful approach; what variables on each axis? Why this shape of graph?	Good reading and interpretation is necessary. Tests modelling skills and transformation of journey formulae

Set B: abstract number	Use for teaching/learning	Use for assessment
Find all the factors of (i) 47 (ii) 48 (iii) 49	Teacher can draw attention to differences in factor structure for these and other positive integers	Correct complete factorisation
What is the same and what is different about the numbers 47, 48, 49?	Teacher could accept a range of characteristics and then draw attention to factors as above, encouraging multiplicative structure as a possible way to think about integers	Not specific about what is being tested. Other same/different distinctions could be made, e.g. odd/even; square or not
Find numbers between 44 and 50 that have: most/fewest factors; even/odd number of factors	Make multiplicative distinctions; teacher could ask for conjectures and check them with 'similar' numbers	Find factors and pick out the requested features
What dimensions of rectangular egg boxes could hold exactly 47, 48 and 49 eggs?	Represent pairs of factors in area format. Since 48 has several answers these could be explored graphically as $pq = 48$ where p and q are factors of 48	Implicitly tests correct factorisation

In England in the 1980s and 1990s, there were many testing initiatives assuming that weaker mathematics students would be able to be more successful with questions that were embedded in contexts. Barry Cooper and Mairead Dunne (1999) provided evidence of class and gender-based inequities created by mathematics test questions dressed up as contextual problems. The solving of such problems assumes that students are capable of flexibility in approaching situational and linguistic differences. Students who had limited outside experience because of socio-economic, cultural or linguistic backgrounds were less able to perform well in such questions. However, Jo Boaler, in her comparative study of two schools with very different teaching methods (1997), noticed that students from one of the schools, when faced with examination questions involving topics they had not specifically been taught, were sometimes able to tackle them by using *ad hoc* problem-solving strategies. Mathematics teaching at that school had largely been a process of working on problems and investigations while developing the skills and habits of inquiry with core mathematical ideas such as linearity, measurement, geometrical reasoning as well as in applied contexts.

Thinking about the usefulness of a context can confuse learning with testing. The former can be much more than practice in application. New knowledge can arise from solving problems, for example the questions in set B of Table 2.1 have the potential to lead an inquiring mind to thoughts about factorisation. Testing regimes are inevitably linked to particular educational aims. Testing regimes are inevitably linked to particular educational aims. If the aims are that students should be able to apply mathematics to realistic contexts then it seems legitimate to test such capabilities but there are assessment implications in the biases that Cooper and Dunne perceived. A written test question can never be solely a 'realistic' context; it always fails to have the relevance and immediacy of actual contextual mathematical capability and also assumes generic awareness of features of the context. Possibly the difference between Boaler's observation and Cooper and Dunne's study is that the students in Boaler's study had been 'schooled' into seeing mathematics as a problem-solving endeavour rather than about rapid application of methods. Teaching that always offers the necessary tools followed by an application of them does not prepare students well for dealing with unfamiliar mathematical structures in unfamiliar situations. This discussion exemplifies how different teaching and testing regimes can influence different students' views of mathematics, as suggested by Ernest.

In contrast to the journey questions of set A, set B requires students to know what factors are—this is not something they might meet in contexts other than mathematics lessons. Unlike set A questions, students cannot get started by using everyday knowledge alone. It helps to recognise multiples and square numbers, but the questions are also accessible to those who have to trudge through various 'times tables' reconstructions or serial memories to find factors. In this latter case, the trudging is likely to be the overwhelming experience, where in the former case the different structures of factorisation are more likely to be observed and exploration of generalisations of such structures could follow. This kind of work is also emancipatory, but not in terms of developing power in the world as described by Margaret Brown (1999) since it is exploration within the imagined abstract world of factorisation. The truth of what is found is subject only to mathematical rules for validity, rather than the realistic truths of set A; the excitement for what is found comes from the power of the mind and the feeling of personal intellectual competence. However, having knowledge yourself, in this case knowledge of factorisation, when it gives access to important facts, procedures and interpretations, is also emancipatory if you cannot buy it, have no way to borrow it, but need to use it.

Antonio Gramsci was Marxist thinker in Italy in the 1920s and 1930s when fascism was on the rise. A period in prison enabled him to reflect on how one of the ingredients of fascist domination was an education system in which the political elite were able to access 'powerful knowledge' while others were locked into having only vocational knowledge or otherwise being given information in an uncritical environment. He saw that people who might otherwise see the mechanisms of unfairness in their own country were instead being steered towards seeing other countries as their enemy rather than seeing and understanding the manipulations of their own leaders. He included, in his analysis of blame, that educational methods often left students without the disciplinary tools of inquiry and critique, while appearing to be successful by only testing those aspects that were seen fit and unthreatening. Some applications of the so-called child-centred theories of Jean-Jacques Rousseau and Maria Montessori were the target of his criticisms because, while they might allow young students to explore the world for themselves, they did not ensure that all students would have access to a core canon of ideas and concepts. He believed that such curricula could prevent students accessing knowledge that could be crucial for future economic, social and political success (Mayo 2014). Educators should 'resist the tendency to render easy that which cannot become easy without being distorted' (Gramsci and Hoare 1971 p. 43). All education (as opposed to instruction[1]) should develop a student's ability to study effortfully and critically in the realm of ideas: 'To reason, to think abstractly and schematically, while remaining able to plunge back from abstraction into real and immediate life' (ibid. p.38).

While I find Gramsci's arguments mainly compelling, they remain aims of education rather than methods. For example, he underestimates the cognitive challenges of early learning, assuming that arithmetic can be learnt procedurally as preparation for later abstract thinking. But the development of mathematical ideas can start from material phenomena (such as playing with physical objects) leading towards being labelled and symbolised as abstractions, or alternatively might start with an abstract idea (such as 'more than') and lead towards insight into material phenomena. For very young children the physical world is the starting point for learning.

As I write Gramsci's concerns are issues in educational debate. Some take the view that a curriculum valuing exploratory and ad hoc approaches can, with suitable pedagogy, nurture powerful mathematical capabilities. Others emphasise that the school's role is to provide knowledge that would otherwise not be available for all. These views are not, in my view, incompatible but the latter seems to have fed into a renewed passion for various versions of 'direct instruction' rather than other routes to powerful knowledge, such as nurturing habits of personal sense-making giving well-designed support for the process. A 'core curriculum for all'

---

[1] I (as did Gramsci) find the word 'instruction' limiting and prefer broader language that is common in Europe, teaching, education, pedagogy and so on, that includes non-content aspects of teacher–student relationships and breadth beyond giving instructions.

argument was behind the original NC but was diluted in practice by the expectation of a wide range of achievement, which became a self-fulfilling prophecy in mathematics by the provision of different tracks for different students. The argument has been renewed by the politicians behind the 2013 National Curriculum. The Secretary of State at the time, though a Tory, even claimed that his vision of a core knowledge-based curriculum was in every sense Gramscian (Gove 2013). But he misunderstood Gramsci. While making similar-sounding criticisms about educational methods that did not sufficiently direct learning towards important ideas he failed to grasp that Gramsci was arguing for an education that would form critical intellect, not an education that would fill heads with a collection of facts that had been selected by the dominant elite. The effect of a conservative, in all meanings, implementation of a curriculum full of knowledge can be to constrain it to what people who are already powerful claim to know and value, rather than to develop the personal power that arises from exercising the structures and uses of knowledge.

Key Differences in Mathematics Education

There are differences in teachers' beliefs that are manifested in their teaching. As Ernest described, beliefs about individual capabilities, the nature of mathematics, the purpose of education, and the role of language vary.

Some teachers aim to affect personal change in individual capabilities as part of their role and others accept personal traits and learning habits as `given'. Thanks to the work of Reuven Feuerstein and others (e.g. Feuerstein et al. 1985; Dweck 2000) it is known that intellectual capability is malleable but often conditioned by background, past experience and educational methods. These differences would lead to various approaches in teaching students who have had some kind of difficulty in learning mathematics. It might be seen as reasonable to simplify the mathematics, slow down the pace of the curriculum, and offer supportive materials and other aids, or even to trim the curriculum to only those aspects of the subject that are necessary for everyday life. Alternatively, it might seem reasonable to improve the student's capability to deal with abstract and multistage ideas, as in Reuven Feuerstein's Mediated Learning system (Feuerstein et al. 1985) or by diagnosing critical gaps in knowledge and 'filling' them, as in various recovery programmes such as Every Child Counts (ECC).

Views of the nature of mathematics differ; is it a collection of procedural tools or a way of thinking about quantitative, spatial and abstract relationships? This contrast shows itself in teaching and assessment methods, and in teachers' own experience. Most often it is manifested in the different values and weightings given to mathematical procedures or relationships, either explicitly or implicitly, through tasks. There is a common view among scientists and engineers that mathematics is the way that the relationships between quantitative variables in their worlds are brought into the manipulable world as a service. While this view often degrades the abstract nature of mathematics, particularly in Science, Technology, Engineering and Mathematics (STEM) initiatives, it supports the notion that focusing on the structure, meaning and manipulations of relationships could be the focus for school mathematics, particularly now that the calculations themselves can be done by machine.

Work in STEM contexts can often be completed with situationally specific methods that do not transfer easily to other contexts. For example, consider this problem:

> In a mine, various lengths of pitprop have to be moved along a tunnel 2m. high and 2m. wide which has a right-angled bend in it. What is the longest length that can be manipulated in this way?

The question is only moderately realistic as it has a closed answer, making the work easy to assess, but many real problems are less defined. Second, this problem can be completed using trial and adjustment (maybe using actual pitprops or models) or measuring scaled diagrams using existing knowledge. The method of solving such problems with a scale drawing might transfer to adaptations using a different angle, or some kind of curve, but the particular mathematical solution itself might not. For an algorithmic method that can then cope with different curves, heights and angles a mathematical model of some sophistication would be required but solving the original problem does not require such a model. Discussions about teaching practices, assessment styles and desired learning outcomes show contrasts between those who see mathematics as generated in and for a specific situation and those who expect it to be transferred without difficulty to other situations; the difference is most marked in cases where assessment styles do not relate closely to usual teaching aims. While getting a situated solution demonstrates the power of ad hoc approaches, the future exercise of power is limited. The *Realistic Mathematics Education* project initiated by Hans Freudenthal in The Netherlands approached this issue by recognising horizontal mathematisation in which several imaginable, concrete, situations that depended on the same underlying mathematical idea would be the focus of classroom work that could eventually lead to vertical mathematisation, in which the underlying structure would be abstracted, itself then becoming the focus of further study and use (Dickinson et al. 2011). Ad hoc methods might be emancipatory, depending as they do on working with an imaginable situation and bringing everyday reasoning to bear on the problem as well as past mathematical knowledge, but the mathematical idea rather than the problem solution is the learning goal.

Education can be seen as being for development of individuals, or for the development of a group, or the future economy, or combinations of all these. Many schools have mottoes that are about the potential of individuals such as: 'be the best you can be' or 'become citizens who contribute to the future' yet the pressures of the examination system can convey that good grades are the most meaningful outcome. Governments prefer to think about the future economy and compare international test results to try to find a link between schooling to successful economic growth. Some schools have happiness or mental well-being in overtly stated aims, but many more might recognise the fundamental importance of these aspects of young people's lives. Contrasts in teaching and assessment practices reflect this confusion of purposes. Most marked is where schools expect students to have different assessment outcomes, a meritocratic approach, rather than describing successful teaching as 'everyone understands' core ideas—a possible interpretation of a mastery approach. As well as illustrating differences between psychological, individualistic or social approaches to education, I also question whether individual emancipation is more likely through individual or group achievement domains. A focus on outside testing also implies a one-size-fits-all agreement about what the core ideas might be, and how they should be sequenced.

Another difference is whether thought precedes or follows language development, or whether they develop together. This contrast is manifested through the roles that language plays in teaching, learning and assessment, and how these roles are deliberately developed in lessons. Again, here is a difference between the individual and social domains of learning; do ideas develop somehow in my own head, or somehow in the classroom as a whole, or in the interactions between me and the teacher? Taking the word 'equation' as an example: does my understanding of equations grow because I know the word and notice it being used in a variety of ways, or do I notice a variety of uses of the equals sign so that when I hear the word 'equation' I know what to focus on? I have deliberately chosen a mathematical idea for which the word 'equation' might not always be used, and for which there are related concepts, such as missing number problems, formulae, or functions, or identities that might sometimes look like equations. For naming a concept or class of mathematical objects 'the meaning of a word is its use in the language' (Wittgenstein 1958, PI 43). However, language does much more than name objects, it also provides structures which can describe thinking so it is possible for different language groups to use different expressions of reasoning. Some teachers in England take this on board in ways far beyond the classical logical constructions of 'if… then …. because' and into mathematical riffs on which mathematical knowledge can be built, such as 'if five is the whole, then two is two-fifths of the whole'; 'when adding tens to a number, the units digit stays the same', ' if two things are equal to the same thing, then they are equal to each other'; knowledge at a variety of grain sizes can be emphasised in this way. There also has to be experience of retrieving and using such structures appropriately. Students might chant 'vertically opposite angles are equal' but be unable to recognise such pairs of angles or know how to use the chanted fact (Watson 2002).

In this chapter, I have presented differences and tensions that influence mathematics education among:

- the purposes and aims of education;
- the nature of school mathematics;
- the malleability of learning capabilities;
- the roles of language.

In later chapters, there will be some descriptions of mathematics pedagogy in a range of contexts and cultures that illustrate how some positions in these aspects can be enacted in classrooms. The next three chapters present more about what is entailed in learning school mathematics.

References

Boaler, J. (1997). *Experiencing school mathematics: Teaching styles, sex and setting*. Buckingham: Open University Press.

Brown, M. (1999). One mathematics for all? In C. Hoyles, C. Morgan, & G. Woodhouse (Eds.), *Rethinking the mathematics curriculum* (pp. 78–89). London: Routledge.

Cooper, B., & Dunne, M. (1999). *Assessing children's mathematical knowledge: Social class, sex, and problem-solving*. London: McGraw-Hill Education (UK).

Dickinson, P., Hough, S., Searle, J., & Barmby, P. (2011). Evaluating the impact of a Realistic Mathematics Education project in secondary schools. *Proceedings of the British Society for Research into Learning Mathematics, 31*(3), 47–52.

Dweck, C. S. (2000). *Self-theories: Their role in motivation, personality, and development*. Hove: Psychology Press.

Ernest, P. (1991). *The philosophy of mathematics education*. London: Falmer Press.

Feuerstein, R., Hoffman, M., Rand, Y., Jensen, M., Tzuriel, D., & Hoffmann, D. (1985). Learning to learn: Mediated learning experiences and instrumental enrichment. *Special Services in the Schools, 3*(1–2), 49–82.

Gove, M. (2013). *The progressive betrayal*. http://www.gov.uk/government/speeches/michael-gove-speaks-about-the-importance-of-teaching. Accessed 11 Apr 2020.

Gramsci, A., & Hoare, Q. (1971). *Selections from the prison notebooks*. London: Lawrence & Wishart.

Mayo, P. (2014). Gramsci and the politics of education. *Capital & Class, 38*(2), 385–398.

Watson, A. (1999). Paradigmatic conflicts in informal mathematics assessment as sources of social inequity. *Educational Review, 51*(2), 105–115.

Watson, A. (2002). Uses of unison responses in mathematics classrooms. *Research in Mathematics Education, 4*(1), 35–49.

Wittgenstein, L. (1958). *Philosophical investigations* (G. E. M. Anscombe, Trans.). Hoboken: Wiley & sons.

3

The Cognitive Work of Learning Mathematics

The Complex Nature of Human Learning

Given that there are different views about what constitutes mathematical knowledge and how it is learnt, any book about alternatives needs to include some of the debated ideas that shape teaching and the expectations on teachers and their students.

David Geary (1995) distinguishes between primary knowledge, which humans have evolved to exercise, and secondary knowledge which is what humans have to learn formally and deliberately. Most mathematics is usually seen as biologically 'secondary' in that it is learnt formally; it is not a feature of natural evolution, and hence there are limitations to what can be perceived and handled at any moment in time. This is the standpoint of the proponents of cognitive load theory, for whom learning is seen as change in long-term memory. In formal education, this is translated into a need to manage teaching input so that particular changes in long-term memory are likely to take place. But in non-school learning contexts any differences between primary and secondary knowledge are less cut and dried, and learning often is more about coming to know in certain ways or becoming a person who acts in certain ways.

A. Watson, *Care in Mathematics Education*, Palgrave Studies in Alternative Education,
https://doi.org/10.1007/978-3-030-64114-6_3

Humans have not evolved to use mobile phones to the extent that we do—yet we learn within settings that give us many things to coordinate. Automatisation is achieved quickly through repeated meaningful use, not through formal purposeless practice of clicking procedures. We do not learn this by being given a phone with only a few buttons to touch; we learn by sorting out a complex situation through motivation, social need, communication, past experience, trial and adjustment, evaluation of the effects of our actions, setting personal goals and recognising when we have achieved them—including asking to be shown what to do. We also learn through random experimentation and evaluation of outcomes; this can lead to different people using different routines to achieve the same results. We manage the cognitive load the new phone presents because, unless we are complete novices, we have general experience of what they do and how we want to use them.

Similarly, humans have not evolved to do skateboard tricks. Although there are starting skills of being able to balance and move on a skateboard at all, learning a trick is not broken down into small steps of progress and presented by authorities as sequences of these. Even the simplest moves require several simultaneous elements. Learning is by observing, imitating, testing, feeling and adjusting. Occasionally, it is possible for someone to use words, or to slow down and show one element, but the main way of learning in this context is imitation of complex behaviour in largely informal, often collaborative groups where peer approval and feedback contribute to personal prowess. A language has developed to describe moves and their relative difficulty and online videos have augmented the available learning methods in recent years. Belonging to the culture, online or physically present, in which this form of learning and its associated language are the norm is a component of learning.

In both of these contexts, there is knowledge that, in David Geary's terms, would be regarded as secondary being neither hard-wired nor evolutionary, but is accessed through our biologically primary functions of sensory perception, interacting with materials, devising routines, communicating with others, and using social and emotional resources to sustain our efforts. It is also significant that this kind of learning happens in settings where the knowledge and skills are immediately useful for some purpose. Thus, some argue, we learn naturally in the context of relevant activity; mathematics might be learnt best in the context of doing mathematics. There are examples of this in Chaps. 6, 7 and 8. George Ellis (2008) elaborates this idea by observing that, in evolutionary terms, humankind develops skills and tools in meaningful environmental contexts. Therefore, he says, developing tools in meaningful contexts is how our brains must work. Our primary knowledge is also largely emotional: the need to know and understand patterns of behaviour; socially based motivation with peers and adult role models; defence mechanisms; creativity; and caring are common to all humans so also must be primary. The challenge to this view is that learning might be more efficient if input is organised to optimise the use of time and effort: learning on mobiles with fewer buttons maybe or practising certain muscle actions before stringing them together.

Cognitive science looks at learning mathematics as acquisition of information by reading, listening, observing what people do, imitating, and observing phenomena. It is broadly agreed that we learn by accumulating familiar schemes and these make future-related learning possible (Paas and Sweller 2011). The fact that we all accumulate vast amounts of knowledge outside classrooms with little apparent effort provides evidence for these natural processes. The role of education is to organise this process towards a culturally sanctioned canon of knowledge, which might be contested, and that canon would include what Vygotsky called the 'scientific knowledge' (Vygotsky 1987 p.27). Scientific knowledge consists of the structured ideas, often abstract, that are only learnt from interactions with knowledgeable others who can introduce us to structures, language, particular modes of inquiry, formal reasoning and symbol systems. However, overlaid onto these theories is the idea that our working memories are limited in how much information we can handle at any one time and its duration. Duration has been variously estimated to be about ten seconds although some people prefer to talk about function rather than duration. Its function is to receive and note incoming information, but processing capacity is limited depending on whether the recipient sees these as groups, structures, or individual elements. The number of elements to be perceived, their interrelationships and the requirements of a mathematical task constitute a 'cognitive load' and this varies according to prior experience (Chen et al. 2015).

Fred Paas and John Sweller's review of 20 years of research and theorising describes the ever-evolving theory of cognitive load and its gradual expansion to include a more complex understanding of the realities of learning situations (2011). I confine myself to considering what is of relevance for mathematics learning during school years.

Memory

Many teachers, pundits and schools adopt aspects of cognitive load theory as a tool to improve teaching and learning. Roughly stated, the main idea being used is that giving students too many things to think about at the same time can be confusing and obstruct learning the target knowledge. Teachers should, they say, therefore organise knowledge into a drip feed of small pieces of information that can be well learnt and accumulated to make a bigger body of knowledge. It is worth noting that even very young children learn in the midst of a very complex world with many stimuli, but while they are learning what they need for day-to-day living they are not necessarily learning structured knowledge systems such as precise language patterns or the counting system without special tasks and an adult who directs their attention. They are, however, learning how to operate in a complex world. Verbal, visual and other sensory capabilities combine for selective perception. Opponents of the 'small pieces' idea as a guide to teaching argue that sorting out complexity is an essential aspect of learning and needs to happen early in teaching so that the new idea has meaning in the context of more complex ideas. Purpose can be understood and habits of mathematical thinking exercised with meaning. However, I recently saw a first lesson in algebra that included definitions of: expression, equation, variable and coefficient, and three different words for subtraction within the first ten minutes. I imagine most teachers would agree that is too much when presented as a string of new words and facts, as it was, without a break. By contrast, Bob Moses (2014) uses almost the whole of a keynote presentation to show how comparing the heights of two people in everyday language can lead to a need to express relationships between unequal heights algebraically; this one idea developed slowly through the reasoning of the audience. Thus, he illustrated how formal understanding of algebraic notation symbolises relationships between quantities that can be 'seen' informally. Short-term efficiency, in his presentation, was subordinated to long-term depth of understanding.

A long-term memory storage system accumulates knowledge, according to cognitive load theory. In education, retention has to be managed, for example by ensuring that knowledge is revisited regularly so that students are reminded of it and can retrieve it easily when required, enriching links between working and long-term memory. But repetition can be about general patterns rather than only particular cases, particularly if representations are kept constant. For example, rather than remembering that $10 - 2 = 8$ as a discrete number fact the focus could be on remembering three ways of representing any additive relationship: $a + b = c$ implies that $c - b = a$ and $c - a = b$. This makes knowing number bonds rather more important than rehearsing subtraction methods in simple cases.

A complex metaphor for long-term memory is that our brains naturally categorise, organise and store information and experiences as schema, not as isolated facts, and these support our current mental actions. The route into memory needs first of all to be through the same 'pathways' in which it was constructed so that schema can be brought into consciousness (Schwartz and Reisberg 1991), so this requires some consistency of images, language, examples, diagrams as more and more pathways and connections are being made. Furthermore, this means that learning facts as serial recitation rather than connected to related meanings makes recall harder as the series has to be thought through to find the desired fact rather than following connections that relate directly to the situation in which the fact is needed. Changes of teacher while memory for a mathematical idea is being constructed can interfere with the process of developing organised and connected memories. Developing long-term memory is not about storing what went into it in the same form as it went in; it is mainly about identifying and storing patterns, relationships, connections and generalities, all of which are the products of our active minds. Richness of knowledge need not be about storing quantities of unconnected facts, but about constructing richly connected webs of meaning.

There are no known limits to long-term memory, and its use expands our capability to manage complex current tasks. A simplistic argument would say that therefore it is important to work hard at memorising so that the basis for learning new ideas is strong, but strangely much of the research on brain function associated with memory is not, in general, about the useful task of schema building but about the less useful tasks of memorising laboratory activities such as learning lists of numbers, or new facts unconnected with other study. It is probably by knowing the structures of mathematics that you have been able to engage with 10 − 2 = 8 above, not by recalling all number bonds to ten and searching for the relevant fact.

There are research results about the widespread practice of offering worked examples; performance on new mathematical problems is sometimes found to be better when students have first seen models of solutions of similar problems, but other studies suggest that tackling something yourself before seeing a solution method leads to better learning, perhaps because it creates a purpose for the method, a need to learn. Also, it is known that if the underlying ideas are not new students no longer benefit so much from model solutions (Chen et al. 2015). Ouhao Chen and his colleagues found that students who already have some relevant knowledge and experience benefit by generating their own solutions first, and all students faced with relatively simple tasks similarly benefit. It is only for absolute novices and complex tasks that seeing worked examples first is best for their own performance on similar tasks. Readers can examine their own experience in general, if not in mathematics, to check that all these results might be true in different circumstances. Furthermore, bear in mind that not all mathematics involves following and replicating worked examples.

Contemplation and manipulation, rather than rushing to the next idea, are known to be as important as practising routines, even when learning procedures (Cole and Cole 2001). In a World Bank study of effective teaching for disadvantaged students the following kinds of practice and repetition were found to be valuable:

- discussion;
- generating stories with the information;
- tying new information to known facts;

- reasoning and explaining how various conclusions were derived;
- making up questions;
- making designs or sketches;
- expressing the concepts in writing or diagrams or physical actions;
- homework that focuses on meaning;
- using imagery;
- peer tutoring that elaborates ideas (Abadzi 2006, p. 179).

In mathematics, it is possible to interpret 'practice' to mean repetitive performance of similar procedures rather than the elaborative actions in this list, although a well-designed set of similar procedures can lead to insights about mathematical structure (Watson and Mason 2006).

Storage metaphors for memory have been around since the 1960s, and possibly before, so it is remarkable that they have become newly fashionable recently, albeit in a more sophisticated form. One reason could be that they describe obvious experiences of learning—surely most people recognise that 'too much information confuses'; 'regular use supports memory'; 'this is boring because I can do this stuff already' and 'show me how to do it so that I can do it'. Perhaps having research results as support brings these out into the open so that teachers' own experience of learning can be better understood and expressed in a shared language. If they think they found learning difficult or repetitive or boring it could be because their own teachers did not use these ideas creatively in their teaching rather than because they themselves were weak students. Applied to mathematics, however, such ideas can limit learning. They are only optimally relevant to the storage of facts and procedures, and much of school mathematics is not about recalling facts or procedures but recognising (or even creating) patterns, concepts and relationships, manipulating their symbolised versions, and becoming the sort of student who does this. In other words, learning is not merely about storage of given facts and methods, it is about how mathematical ideas are related, transformed, connected and represented in long-term memory. Moreover, in mathematics the new ideas that students have to learn are never completely new, they always build on what has gone before.

Learners are never novice students in the everyday meaning of the word, they always have prior experience of the materials or symbols in a problem and the usual expectations of how to learn in the classroom, such as observing similar and different features of examples, seeking and using patterns, spotting relationships and so on. However, in cognitive load theory the word 'novice' applies only to someone for whom the target fact or procedure is absolutely new, but many may have an existing store of related knowledge, experience and ways of behaving that will help them. As well as having a fund of knowledge they also operate in the social and emotional context of school while the all-embracing business of growing up goes on. Pure application in teaching of research results, particularly those arising in laboratories, takes little of this into account.

Memory for facts can be of different kinds: episodic (the fact and its circumstances), procedural (a list of things to do) and semantic (a representation of meaning) (Baddeley et al. 2015) with emotional, social and metacognitive dimensions (Damasio 1999; Roebers 2017). Episodic and semantic memory are different, and many teachers try to combine the two by linking semantic memory of facts to events, environment, materials, conversations. They might ask: 'Do you remember when we?' for remembering the original teaching event (Tulving 2002).

The content of memory matters of course. Different concept images will be constructed by different people with different teachers, resources, and language patterns as well as because of individual differences in attention, listening, and prior images and schemas. Jennie Grammer, Jennifer Coffman and Peter Ornstein (2013) note that those students whose mathematics teachers used an extensive range of methods to aid memory (associations, patterns, images, rhythms, etc.) were significantly more able to complete sorting, classifying, ordering and grouping tasks than those whose teachers used less variety.

Chunking

Cognitive science has something useful to say that addresses mathematics in a more complex way and also explains why, as children grow older, they can deal with more complex concepts. This is the idea that 'chunking' knowledge connects what previously might have been several separate items of knowledge into one item which can, therefore, be handled within working memory limitations where before it might have been too much to handle. A natural propensity to generalise supports this and language and symbolism play a part. For example, '7' has several properties of number, numerosity, structure and magnitude but it is by its context that anything can be done with it. Different conceptual features of '7' are needed for each of the following tasks:

What counting number comes after 6?
I have 8 sweets and eat one of them, how many do I have left?
On a numberline, what is half-way between 5 and 9?
$3 = 21 \div ?$
What is the missing number in this sequence, and why: 2, 3, 5, … , 11, 13,17,19?

All these possible relationships could be connected to the word and symbol for seven, and also each meaning is connected to various generalities for small positive integers, which might later be connected to a set of wider generalities about real numbers. A 'subtracting integers' chunk is not much use if a number task involves a 'prime number' chunk. David Tall and Shlomo VInner developed the idea of 'concept images' (1981). These are collections of properties, behaviours, representations, relationships and techniques that are clustered around a mathematical concept. These clusters will have their own internal organisation which reflects an individual student's learning experiences. '7', hopefully, is not a separate number but would be included in the image of small positive integers and connected to images for odd numbers, primes and so on. It may contribute to the formation of concepts through sharing properties with other numbers; concept images for even and odd numbers, consecutive numbers, prime numbers and so on can develop in long-term memory through experiences with particular numbers. Chunking is an embedded feature of mathematical structure, and that is what generalisations, definitions and symbols are for.

Knowledge of mathematics is not merely an accumulation of information but also includes the ways in which we select and coordinate mathematical actions. This is done differently in mathematics than selection and coordination in, say, cooking or language learning. In cookery learning we might imagine a taste, a smell, a process and anticipate our need for tools; in language learning we might focus on how to use our lips, tongue and so on to say the word or try to grasp its meaning and hence its potential use. In mathematics, a word or symbol or situation might bring a statement, a set of examples, a diagram to mind. Mathematical selection and coordination needs to be experienced with chunks, or concept images, of mathematical ideas.

In most of school, mathematics chunking is associated with representation. For example, a coordinate graph of $y = x$ chunks a relationship of equality between two variables, one of which might be dependent on the other; the value of y might depend on the value of x or vice versa. After some experience, such a graph can be presented to students without overloading or confusing them but as something they recognise, can use and interpret. However, such chunking has a flip side which is that this particular graph is often presented as a static picture with equal aspect axes, so the idea that $y = x$ is a diagonal of a squarish picture, and makes 45° with the horizontal and vertical axes can become chunked along with everything else—so much so that any straight-line graph that is at 45° with the horizontal through the original might be assumed to be a representation of $y = x$. So while chunking makes it possible to deal with more and more complex ideas, it can also lead to assumptions about those ideas that are hard to change. Teaching needs to ensure that the contents and structure of students' concept images contains enough varied examples, the example space (Watson and Mason 2005), so that oversimplification can be avoided and symbols can be read fluently and meaningfully. Chunking does not stop there of course; $y = x$ contributes to higher order chunks of graphs of proportional relationships, linear graphs, functions and so on.

Managing the Load on Memory

If 'problem-solving' is taken to mean 'doing exercises that use a recently learnt method' then studies of immediate cognitive load and memory have a role to play. This limited version of what 'problem solving' means is widely used so that tightly managed teaching, followed by a sequence of similar questions, can be described as efficient for solving problems. However, in Europe the word 'problem' is more generally used to refer to unfamiliar questions that require some generic problem-solving approaches and transformation of relevant knowledge. Success in such tasks involves not only some prior knowledge that might be helpful, but habits of mathematical inquiry, generic problem-solving techniques, and reflection on the processes and progress towards a solution. Enculturating students into using some of these thinking processes throughout their mathematics lessons can convince them that they have some power; there are things they can do when faced with an unfamiliar problem. Regular use of a repertoire of problem-solving processes can promote the adoption and habitual use of mathematical methods of inquiry.

The research literature in this field is vast and varied; there have been many attempts to teach mathematical problem-solving, in its complex sense, by applying a sequence of generic moves in the hope that this will improve students' approach to future mathematical tasks. Frank Lester (1994) reflects that this kind of direct teaching of processes is not enough, there also has to be the development of metacognitive skills to monitor and evaluate progress through the task. Through such monitoring, prior knowledge can be brought to bear effectively on unfamiliar problems. His review of the literature during a period when problem solving was of intense interest in US research shows that students' problem-solving abilities improve if they have plenty of planned experience of solving a variety of problems, over time, and know that this is important (ibid. p.665/666). Moreover, it helps if their prior knowledge is of structures rather than only facts and methods.

Managing cognitive load is sometimes written about and researched as if it is the responsibility of the teacher or textbook to offer an appropriate simple load to students, but to tackle a problematic piece of mathematics always involves managing cognitive load by: interpreting symbols; seeking underlying structure, which will bring past experience of similar and related structures to the fore; selecting suitable strategies and processes, including simplification methods. These could be manipulations, calculations, substitutions, chains of reasoning and logical structures, selected from memory and past experience of similar or related problems using analogical reasoning. It is often helpful to try simple cases, decide the order of actions, and finally to reflect on their effects in order to construct conjectures. These constitute mathematical thinking and manage cognitive load by creating and dealing with subproblems. Descriptions of handling problematic mathematical ideas in these ways, such as those of George Polya (1945), precede articulations of cognitive load theory by decades.

A limited view of 'split attention' (Chandler and Sweller 1992) is sometimes cited by pundits to avoid the load involved in using more than one representation; by contrast the use of two representations, particularly verbal and visual, can support conceptual learning (Paivio 1971; Duval 2006). Bunching connections into concept images is particularly useful in mathematics since the ideas are abstract and always need to be communicated in visual, verbal, symbolic, imaginary or material form, often necessarily using two or more representation at the same time, such as using spoken and symbolic number along with position on a number-line, or algebraic and graphical representations of functions along with hand gestures. Teaching approaches that offer limited cognitive load and do not exploit these other possibilities for mental action are likely to engender a simplistic view of mathematical activity and dependency on teacher and textbook rather than independent thought; it is agreed that there has to be some cognitive load for any learning to take place at all.

Executive Function

In the late 1980s, scientists became interested in activity in the prefrontal lobes of the brain and began to learn in detail how this part of the brain appears to be activated when people monitor and regulate their activity while doing given tasks. These processes, the executive function of our brains, allow us to operate in the complex world, whatever our present focus. The processes are at work whether I am contemplating '7' or '$y = x$' or what to cook for dinner or even the question 'why are we here?'. Executive function is generally subgrouped into three kinds: working memory (what is attended to right now); inhibition (what is around but ignored right now); and task switching or shifting (using what is known from one focus to with another focus) (Hofmann et al. 2012.) To relate these to doing mathematics I find the survey of literature offered by María Beatriz Jurado and Mónica Rosselli (2007, 213–333) more useful than broad categories. They found over 20 ways of describing functions that could be, or had been, tested using cognitive science research methods. Some of the ways overlap or may be different ways to describe similar facets, or appear to contain others, according to the language used by original researchers. They are of differing grain size and for the purposes of this book I have organised them according to their relevance for learning mathematics, assuming tasks that are most often more complex than being told facts and shown methods (see Table 3.1).

The functions they identify closely resemble experiential accounts of working on mathematical problems, such as those from George Polya (1945), Alan Schoenfeld (1985), John Mason, Leone Burton and Kaye Stacey (1982). In some methods of teaching, applying these generic executive capabilities in all tasks, whether familiar or not, is the norm.

So far little attention has been given within cognitive science to the association of these processes with mathematical learning. Cognitive science is progressing towards having more to say about varieties of mathematical work other than merely working on given examples. Some

Table 3.1 Executive function during complex mathematical tasks (synthesised from Jurado and Rosselli 2007)

Task engagement	Associated executive functions
Deciding and planning	Volition, initiation, task analysis, identifying goals, setting goals, planning, generating strategies, considering visuo-spatial sketchpad and phonological loop
Ongoing in-task functions	Solving novel problems, cognitive flexibility, reasoning, creativity, abstract thinking, problem-solving, concept formation
Self-controlling actions	Cue-directed behaviour, purposeful and coordinated organisation, purposive action, sequencing complex actions, supervising and self-controlling attention, preserving goal-directed action, impulse control, inhibiting distractions, monitoring and controlling strategies
Reflecting on the effects of actions	Adapting goals, altering goal-directed behaviour, concurrent manipulation of information, maintaining effective performance
Post-task	Analysis of success of strategies, concept formation

findings so far have suffered from the fact that the research tasks have often been unrealistic in terms of normal mathematics school learning and not carried out on students in school settings. The purpose of the tasks used in cognitive science research is often to test and develop cognitive theory against constrained tasks using a box-model of memory or magnetic imaging of brain activity. Where school students in normal classroom settings have been the subjects, research tends to be on the effects of short-term interventions using reseasrch tasks rather than medium or long-term teaching of a holistic curriculum with the usual teacher. The purpose of tasks used for teaching is often to develop the long-term knowledge and generic mathematical capabilities of students rather than retention of one item of knowledge.

Cognitive science can rarely be regarded as a guide to teaching because that is not its role, although cognitive load theory is progressing in that direction. What is often missing is the complexity of classrooms and students as social beings. Neither can it tell us about what can be achieved by pedagogic interactions of the kind Lev Vygotsky talks about, where knowledge, ideas and thinking capabilities are expanded by mediation provided by suitable tasks, talk, tools, including digital tools, or specially designed materials (Daniels 2016). Bodily senses also play a part in mediation since they provide perception of mathematical phenomena. In mathematics education complex processes, and their role in mathematics classrooms, have been a major research focus for several decades and are also present in many teachers' accounts of their practice. There are roles for all areas of research, including looking from a sociological or linguistic perspective, and also for analysing tasks from a purely mathematical perspective for their conceptual affordances.

A further ingredient in thinking about the cognitive challenges in learning mathematics is Reuven Feuerstein's et al. (1985) development of mediated learning, based on intense interactions with individuals or small groups. These show that learners' cognitive capabilities can be expanded through reciprocal, reflexive, responsive interactions with trained adults over time. Michael Shayer, David Johnson and Mundher Adhami (1999) used these principles in the *Cognitive Acceleration in Mathematics Education* project. Their interventions in science and mathematics appeared to expand cognitive capabilities not only in the focus subject but also in other areas of their schoolwork. Feuerstein's approach modelled how to be an actively thinking person in reciprocal dialogue using the language of reasoning.

Care for Mathematical Knowledge and Cognition

Heeding Michael Young's warning that education might over-emphasise memory (2019), there needs to be a balance between application of models of memory and other aspects of mathematical development. Learning needs to include: knowing how and when to act mathematically; having several ways to access, use and restructure memory in mathematically relevant ways; knowing how to work with unfamiliar mathematical situations; knowing how to use mathematical methods and tools to manage cognitive load; knowing how to coordinate knowledge gained from everyday experience with formal, symbolic approaches to mathematics. Without these forms of knowing students are dependent on being given knowledge by others.

Teaching mathematics can take all forms of knowing into account using cognitive science to explain some classroom learning phenomena and affirm or challenge professional knowledge while also providing opportunities to learn those aspects of mathematics that are not solely about memory. Language emerges as an important component of teaching, particularly in shaping knowledge and supporting reasoning.

In the next chapter, I comment on forms of pedagogy and how they relate to mathematical content, knowledge and cognition.

References

Abadzi, H. (2006). *Efficient learning for the poor: Insights from the frontier of cognitive neuroscience*. Washington, DC: World Bank.

Baddeley, A., Eysenck, M., & Anderson, M. (2015). *Memory* (2nd ed.). London/New York: Taylor & Francis.

Chandler, P., & Sweller, J. (1992). The split-attention effect as a factor in the design of instruction. *British Journal of Educational Psychology, 62*(2), 233–246.

Chen, O., Kalyuga, S., & Sweller, J. (2015). The worked example effect, the generation effect, and element interactivity. *Journal of Educational Psychology, 107*(3), 689–704.

Cole, M., & Cole, S. (2001). *The development of children*. New York: Worth Publishers.

Damasio, A. (1999). *The feeling of what happens: Body and emotion in the making of consciousness*. Boston: Houghton Mifflin Harcourt.

Daniels, H. (2016). *Vygotsky and pedagogy*. London: Routledge.

Duval, R. (2006). A cognitive analysis of problems of comprehension in a learning of mathematics. *Educational Studies in Mathematics, 61*(1–2), 103–131.

Ellis, G. (2008). Commentary on 'an evolutionarily informed education science' by David C. Geary. *Educational Psychologist, 43*(4), 206–213.

Feuerstein, R., Hoffman, M., Rand, Y., Jensen, M., Tzuriel, D., Hoffmann, D. (1985). Learning to learn: mediated learning experiences and instrumental enrichment. Special Services in the Schools, 3(1-2), 49–82.

Geary, D. (1995). Reflections of evolution and culture in children's cognition: Implications for mathematical development and instruction. *American Psychologist, 50*(1), 24.

Grammer, J., Coffman, J., & Ornstein, P. (2013). The effect of teachers' memory-relevant language on children's strategy use and knowledge. *Child Development, 84*(6), 1989–2002.

Hofmann, W., Schmeichel, B., & Baddeley, A. (2012). Executive functions and self-regulation. *Trends in Cognitive Sciences, 16*(3), 174–180.

Jurado, M., & Rosselli, M. (2007). The elusive nature of executive functions: A review of our current understanding. *Neuropsychology Review, 17*(3), 213–233.

Lester, F. (1994). Musings about mathematical problem-solving research: 1970–1994. *Journal for Research in Mathematics Education, 25*(6), 660–675.

Mason, J., Burton, L., & Stacey, K. (1982). *Thinking mathematically*. London: Pearson Higher Education.

Moses, R. (2014). www.youtube.com/watch?v=fEYIi1iYFsA. Accessed 11 Apr 2020.

Paas, F., & Sweller, J. (2011). An evolutionary upgrade of cognitive load theory: Using the human motor system and collaboration to support the learning of complex cognitive tasks. *Educational Psychology Review, 24*, 27–45.

Paivio, A. (1971). *Imagery and verbal processes*. New York: Holt, Rinehart & Winston.

Polya, G. (1945). *How to solve it: A new aspect of mathematical method*. Princeton: Princeton University Press.

Roebers, C. (2017). Executive function and metacognition: Towards a unifying framework of cognitive self-regulation. *Developmental Review, 45*, 31–51.

Schoenfeld, A. (1985). *Mathematical problem solving*. New York: Elsevier.

Schwartz, B., & Reisberg, D. (1991). *Learning and memory*. New York: WW Norton & Co.

Shayer, M., Johnson, D., Adhami, M. (1999). Does CAME work?: report on key stage 3 results following the use of the Cognitive Acceleration in Mathematics Education project in year 7 and 8. In *Proceedings of the British Society for Research into Learning Mathematics, 19*(2), 79–84.

Tall, D., & Vinner, S. (1981). Concept image and concept definition in mathematics with particular reference to limits and continuity. *Educational Studies in Mathematics, 12*(2), 151–169.

Tulving, E. (2002). Episodic memory: From mind to brain. *Annual Review of Psychology, 53*, 1–25.

Vygotsky, L. (1987). Thinking and speech. In R. W. Rieber & A. S. Carton (Eds.), *The collected works of L.S. Vygotsky, Volume 1: Problems of general psychology* (N. Minick, Trans.). New York: Plenum.

Watson, A., & Mason, J. (2005). *Mathematics as a constructive activity: Learners generating examples*. Mahwah: Lawrence Erlbaum Associates.

Watson, A., & Mason, J. (2006). Seeing an exercise as a single mathematical object: Using variation to structure sense-making. *Mathematical Thinking and Learning., 8*(2), 91–111.

Young, M. (2019). *Cultural capital and curriculum*. http://ioelondonblog.wordpress.com/2019/09/09/cultural-capital-and-curriculum-will-ofsteds-new-frameworkencourage-better-education-in-our-schools/. Accessed 11 Apr 2020.

4

Content, Pedagogy and Learning States

Worthwhile Knowledge

Four decades ago, Jerome Bruner suggested:

> We might ask, as a criterion for any subject taught in primary school, whether, when fully developed, it is worth an adult's knowing, and whether having known it as a child makes a person a better adult…it should follow that a curriculum ought to be built around the great issues, principles, and values that a society deems worthy of the continual concern of its members (1960 p.52).

This raises the question of what is worthwhile. The previous chapters indicate that there are differences of opinion. For pure mathematicians, understanding mathematical objects and structures, together with precise logical reasoning, might be seen as the most worthwhile forms of knowledge whereas employers might express a desire for flexible application and problem-solving capabilities. Nel Noddings, a leading educational philosopher who used to be a mathematics educator, argues that that only applications of arithmetic, statistics and some spatial reasoning are worthwhile for all:

A. Watson, *Care in Mathematics Education*, Palgrave Studies in Alternative Education,
https://doi.org/10.1007/978-3-030-64114-6_4

> … to have uniformly high expectations for all students in mathematics is morally wrong and pedagogically disastrous. It is part of a sloganised attempt to make our schools look democratic and egalitarian when, in fact, they are systems continually struggling for tighter control (1993, p. 159).

As I write, the aims of the mathematics National Curriculum (NC) for England include fluency, reasoning and problem solving for all ages, but the testing system promotes arithmetical calculation as the most important goal of primary mathematics. An alternative goal might be the deep and usable understanding of fundamental structures, such as the number system, arithmetical operations, and spatial properties. Calculation on its own is not powerful in the sense of giving students the intellectual tools of power; I would argue that what England has currently at primary school level is the opposite, having the effect of constraining learners to do uncritically what machines can do. The strengths of such an approach are that it is universally recognised, easy to test, and is thought to provide a basis for quantitative issues that arise in everyday, employment, other subjects, and further mathematical learning. Even in Gramsci's writings about politically powerful education basic arithmetic was seen as an unproblematic preparation for the critical education that should follow it (Gramsci and Hoare 1971). In fact, it is far from unproblematic as any teacher trying to get everyone to do long division in time for national tests at age 11 knows. It can happen that those who start school with least cultural capital get a restricted experience of number and calculations in order to help them 'catch up' to a compliant level of procedural skill, possibly without the accompanying conceptual depth that allows them to apply it usefully and powerfully.

In the context of educational alternatives, it makes sense to consider whether it is more or less worthwhile to 'upskill' students in the kind of mathematics that is publicly understood to be important, than to develop young minds to exercise their intellect. Does one exclude the other? Is it possible to do both? There will be examples in Chaps. 7 and 8 to show that a temporary period of upskilling has value in some situations within a mainstream context of developing young minds.

Abstraction

One lovely thing about mathematics is that learners' construction of complex ideas from chunking simpler ideas mirrors mathematics as a discipline. Understanding that number can be used both to label individual items and also to describe the total amount of items in a set is an important shift for students. Recognising that angles can be defined by the height they attain from the centre of a unit circle (their sine) is a major concept mid-adolescence. If the gradient of a straight-line is described by using the ratio of sides of a right-angled triangle, then the idea that for a curve such a triangle can be made very small follows, and that very small triangle conveys something about a tiny part of the curve. Human exploration by trying to extend what is known into new places is similar to the historical progress of mathematical ideas, though not at the same slow speed or even in the same order.

I illustrate this with the development of a mathematical concept throughout school: if a variable gets multiplied by something and then something constant added it would be called linear. The structure can be recognised in several different contexts—for example: buying 15 items at 3.99 each and paying an extra 10p for a bag. The 10p is not dependent on the number of items. The same structure is manifested by: skip counting not from zero by repeated addition; buying three same price train tickets online and paying a standard processing fee; taking one packed lunch for each child plus one for the teacher on a school trip. All these involve gradient, or rate per unit, plus a constant starting value. School mathematics includes recognising this relationship when it appears algebraically or graphically: $ax + b$; $c(x + d)$; $p + qx$; and even when the unknown is a function itself: $v(sinx) + w$ is linear in $sinx$. Recognising that the coefficient and constant can be negative is part of knowing the structure. While linear equations and graphs follow this format, in linear algebra they cannot contain constant terms. For a function to be linear in linear algebra these two things have to be true: $af(x) = f(ax)$ and $f(a + b) = f(a) + f(b)$. So what challenges are there for students? Recognition of structure has to include the domain which is currently operating. At least the word linear means that a straight-line would arise from graphing one variable against another—if it goes through the origin it would be a linear function; if it goes through some other point it would give a linear equation with a constant term. So as well as learning what something means students must also appreciate what similar things it does not mean, and also when it means what they think it means and when it means something else. This starts with very young children, when halving something stops meaning 'two parts' or even 'two roughly equal parts' but 'two exactly equal parts that might not look the same'.

A discussion is needed here about the meaning of 'abstraction'. For mathematicians, abstraction means the generalisation and expression of a mathematical concept, such as linearity, that is manifested in certain relations between real world or mathematical objects that have an equivalent structure. The abstraction can be treated as an object through using symbols, even when it has no concrete manifestations. For example, pure numbers, algebraic identities, polynomials, groups, strings are all abstractions. Vadim Krutetskii (1969) studied students who had been labelled as 'mathematically gifted' and found that one of the characteristics they had in common was memory for mathematical structure, not memory for methods and facts. Pedagogy that places importance on structure is associated with better mathematical problem solving (Lester 1994). In the example of linearity above structure is being developed, and then reorganised and recategorised as new mathematical experiences come on board.

In brain science, abstract thinking refers to thinking that is generated by the thinker without perceptual stimuli and can refer to past or future events, or relations between simpler objects of thought that have been stimulated by concrete experiences, such as physical objects. There is some, but the incomplete and potentially confusing overlap between these meanings, since much of what brain science calls abstract would, in mathematics, be represented symbolically and hence be visually present in some form, and some would be described as mathematical imagery.

Assumptions that cognitive load can be micro-managed by teachers in a way that is suitable for all assumes a uniformity of past experience, prior chunking, concept images, personal example spaces, memory for facts and procedures and sounds like a far-fetched idea in most classrooms. Also the superficial use of cognitive science to justify teaching mathematics in small steps to be memorised, as I have explained in Chap. 3, limits mathematics and might also lead to unhelpful teaching that ignores existing knowledge and experience. A further difference is that cognitive scientists tend to talk about efficiency and optimisation as if these are desirable qualities of instruction where educationists tend to see time and resource limitations as obstacles to slow, deep, exploratory learning that is preferable for individual development. These differences, among others, add to the difficulty of using results from brain science to apply to the abstract reasoning that comprises most of pure mathematics, and to mathematical behaviour. Does it matter whether using Cuisenaire rods to represent number is abstract or not? Would imagining Cuisenaire rods be abstract or not? Would remembering an actual calculation done yesterday be abstract or not? The more important question for teachers is—what kind of learning is afforded by these?

Different Pedagogical Stances

How do different pedagogical stances address worthwhileness? Superficial reading of research summaries might suggest that there are two opposing ways to teach mathematics, often referred to as 'direct instruction' versus 'inquiry'. This is a false dichotomy. These are only two positions in a field of possibilities.

The description 'direct instruction' has evolved to have many different meanings, most of which relate to teachers showing, telling and modelling mathematical facts, methods, concepts to students rather than setting up situations for inductive reasoning from exploring examples. A source used by teachers for learning about direct instruction is Barak Rosenshine's *Principles of Instruction* (2012) in which a very broad meaning is given that could include any form of teaching that starts with a review of earlier learning, introduces tasks that have a specific mathematical goal in mind, and gives feedback that synthesises and corrects the students' path towards that goal. Tasks offered by proponents guide learners towards an expected outcome along one reasoning pathway. Research that pits 'direct instruction' against 'discovery' or 'inquiry learning' or some other type of pedagogy varies widely in findings. This is because each comparative study depends on the contexts of the study, the educational aims of both methods, which may differ, the test used for comparison and a range of contextual educational variables. Also any method is subject to local interpretation by teachers. This is perhaps why it is easier to research methods that constrain what the teacher does and says, such as totally scripted or totally no-guidance lessons, than methods that depend on in-the-moment interactions between teachers and students since these will always vary in pace, focus and direction. This, along with different meanings of 'effectiveness', explains why some studies try to compare two precisely defined and controlled methods or merely compare one defined method with undefined 'business as usual' as a control group. A recent review of research (Hodgen et al. 2020) noted a low number of relevant studies about 'student-centred learning', a loosely defined set of methods, but those that existed showed positive results with a high effect size.

The most convincing research about the value of direct instruction, in which programmes of basic skills and facts are taught by teachers in prescribed ways, relates to disadvantaged children and is often about programmes designed to ensure they have the basic fluency in number and reading by the age of about 8 that will support later learning. The argument is that without basic fluency they will be unable to access any powerful knowledge and develop more complex capabilities. Their aim is to bring children from restricted backgrounds to a level of knowledge that enables them to take part alongside others in the reasoning and problem solving of mathematics that fully characterises the subject. Helen Abadzi's study for the World Bank (2006) of effective pedagogy for poor children shows repeatedly that children from middle-class homes, worldwide, receive and benefit from a richer, more conceptual, varied diet of mathematical activity but the argument that poorer children need more direction is made variously on compensatory, developmental, resourcing, economic and equity grounds. In her study, poor children are seen as so different that they need different teaching to take account of delayed early development and lack of resources, skilled teachers, home support, food and housing. The conclusion is that they need the accelerated and directed provision of basic knowledge for a limited period so that they can join in with mainstream soon to benefit from a richer curriculum.

These conclusions are not relevant in some contexts in England, but are relevant for some students in many contexts.

Studies that purport to show that some form of direct instruction is more effective than other methods tend to focus on factual and procedural task performance learnt over a limited time rather than mathematical thinking that might benefit students more in the longer term. However, one strong result is that Siegfried Engelmann's version of direct instruction, DISTAR, was effective over other methods when helping young children who have not kept pace with peers learn basic number facts, number relations and arithmetic (Kim 1972). It is easy to see why in terms of cognitive load: the lessons are scripted with well-designed prescribed examples, coherent conceptual sequencing, forms of questioning, responses to students, formats for positive and useful feedback, and timings all focused on one small learning outcome. It is far more rigid than Rosenshine's principles require. The lessons control experience and sense making in a way that resists diversity and creativity, seeing these as potential hindrances to learning rather than as sources of knowledge. I could describe this as cognitive control. But there is far more at work in the DISTAR model than is usually replicated in 'direct instruction' studies. It is a 'rescue' strategy that aims to return seriously disadvantaged children, such as those who may have delayed language, social or even brain development due to malnutrition, neglect or abuse, back into mainstream schooling and has the following features:

- the full method is carried out with small groups, not whole classes, so teachers can observe and react to what the students were doing and modify pace accordingly;
- the small groups consist of children who have a delay due to having a disadvantaged background;
- the daily lesson involves 15 minutes of direct instruction from a script, led by the teacher, including rapid student responses;
- following direct instruction are 30 minutes of independent work; with a small group and a teaching assistant there is plenty of time for remedial interactions;

- the lessons take place in schools that have committed to having daily lessons with staff and teaching assistants who have been trained to the materials and scripts;
- schools have health, nutrition, dental and social work support for the DISTAR cohort;
- schools organise the timetable, rooms and staffing so that these lessons with small groups are possible;
- parents and carers are included in the training and have a knowledgeable supportive role in the home and sometimes in school too, so when children are asked to 'think at home' this becomes a family task;
- the whole experience is over a limited timespan of one or two years.

What is not in this list is the nature of the scripts and the sequencing of the concepts. As with all research about mathematics teaching, it matters what the content is and how it relates to prior and future knowledge. But DISTAR is not a 'do-it-yourself' system, nor a 'select-from-a-website' system. It is a 'do precisely this and students will learn' system for relatively short-term use to enable delayed students to catch up with others whose lessons might include more games, puzzles, applications, 'say what you think' tasks and so on. DISTAR cannot be taken to apply to mathematics students in whole classes without the additional social, cultural and physical components. It does not provide the breadth of a full mathematical experience, but Engelmann's own teaching does provide flexible knowledge of number operation structures (Engelmann 2013). DISTAR provides wraparound care for the whole student. Far from being mainstream mathematics teaching, DISTAR creates an alternative educational space in which students flourish while learning some socially recognisable and mathematically essential processes and facts. As such it gives many clues about what might be possible or even necessary in alternative educational spaces, namely care for individual students as living beings with outside lives and available personalised help within a cognitively well-structured learning context. However, not all studies of its effects outside the work of the originators showed superiority over other methods (e.g. Kaufman 1973). A common variation in roll out is loosening the connection between the cognitive control aspects and social care aspects.

Unfortunately, the research results about the value of such constrained direct instruction are often applied more broadly to apply to all students at all ages in all aspects of mathematics so that the development of a range of mathematical knowledge and thinking is restricted to what is directly told or shown by the teacher or materials. Thus, learners who may not need cognitive control are subjected to it in the name of educational efficiency and optimisation. Clearly, from the example of linearity above, memory alone is not enough for robust mathematical learning because memorised meanings may need modification, the processes that Jean Piaget described as schema development in which students assimilate new ideas alongside old ideas, then accommodate them into a new schema by adjustment and restructuring (1936). Because many teachers know that the processes described by Piaget are not the same for all students and all mathematics, pedagogies other than 'being told or shown one pathway' are often augmented or replaced by hybrid learning spaces in which the activity is about doing, exploring, discussing and using mathematics—variously described as problem-solving, inquiry, student-centred or investigative methods.

In a synthesis of comparative studies Ard Lazonder and Ruth Harmsen (2016) confirm variability in the interpretation of 'inquiry'. Most well-described studies they found were in science with a few in mathematics; comparing inquiry teaching to more direct methods showed that success was dependent on the amount and nature of guidance given to students before and during inquiry. Absence of guidance was not a useful practice, but on the other hand a great deal of guidance turned the inquiry into following a procedure, defeating the potential to develop students' own capabilities. Nevertheless, even with this variety, research supports the idea that students are sometimes more likely to remember and understand ideas when they have been involved in their generation (Chen et al. 2015). Also, engagement is stronger when learning has involved overcoming obstacles, and also when there have been moments of unexpected insight (Abadzi 2006, p.175). These are not experiences that can easily be provided when teachers totally control both input and the nature of acceptable output, although controlling input can sometimes provide the raw material for a particular insight.

Why do different students in the same context learn different things? This question gave rise to the development of Variation Theory in which Ference Marton (2014) and his colleagues realised that learning starts with sensory perception; we notice what varies against an invariant background. The implication is that teachers need to be careful about what variations are offered to students, and in what order. For students to notice particular relationships or patterns of behaviour they need to have attention directed by controlled variation that juxtaposes differences and similarities. Similar ideas have developed in: the Chinese concept of *bianshi* which is the presentation of differences among examples and non-examples; Zoltan Dienes' (1969) distinctions between perceptual variability (the variations we perceive) and mathematical variability (variations inherent in mathematics); Caleb Gattegno's presentations of structured examples to promote conceptualisation (1961); and Engelmann's pedagogic principles for classifying qualities through variability, and constructing relationships through similarities within a class (Engelmann and Carnine 1982). Attention and perception are at the heart of these ideas, so language, representations and gestures are important pedagogical ingredients. Most of what has to be learnt in school mathematics are not facts and methods but relationships in which some elements depend on other elements. Students need help to 'see' and make connections that are not directly perceivable. A lesson using the same materials and the same examples can be very different if gestures and speech vary (Kullberg et al. 2014). Understanding gesture is a natural act, and understanding language also seems to develop naturally, so these do not add to cognitive load but instead support and extend learners' capacity to handle ideas. The intended object of learning is enacted by the teacher through materials and actions as described earlier, but even so the lived experiences of learners can differ for a variety of personal and environmental reasons. Thus, the journey of a mathematical idea from what was intended to what is experienced by a learner can only ever be partly controlled, even by the most carefully scripted and choreographed teacher.

To illustrate how different guidance can generate different learning it seems the right time to revisit the factorising questions in Chap. 2:

Find all the factors of (i) 47 (ii) 48 (iii) 49

What is the same and what is different about the numbers 47, 48, 49?
Find numbers between 44 and 50 that have: most/fewest factors; even/odd number of factors
What dimensions of rectangular egg boxes could hold exactly 47, 48 and 49 eggs?

The second version is goal-free and, according to Paas and Sweller (2011), this presents a fairly light cognitive load because there is no need to keep checking whether you are approaching the target knowledge or result, and no clues about the intended mathematical purpose of the task. The first version is heavily guided, but it is not clear what you are supposed to learn from the experience of completing the task. The third and fourth versions are guided and the third one suggests there is something to notice about primes and squares and maybe something about other multiples too. There are no prompts to make conjectures and test them with other numbers. What matters is not what these various approaches are called but what learning is anticipated, what happens to the results of students' thinking, what verbal guidance is given, whether students work together and share ideas or work in isolation, whether there is any review or closure developed from the work and how it relates to what went before this task and what follows on from it. The second might have a lighter cognitive load—so light that little can be learnt without further teacher input; the others consist of steps that could lead towards a conceptual insight, but may need prompts from a teacher to remind them about mathematical habits such as sorting, classifying, making conjectures about relationships, generalising and even justifying. For these to be habits students need to have frequent opportunities to develop them, and for them to be openly valued and understood as mathematical methods of inquiry. Cognitive load might be low for some stages of this, higher for others, but any measure of it from the interacting elements in the task would be overshadowed in classrooms by the social and emotional engagement, learning habits and recording abilities of students, and the ways of behaving that are usual in that teacher's lessons.

Beyond Cognition to Learning States

The storage model of memory does not include all experiences of 'coming to know'. It defines learning in its own terms as processing knowledge-objects to be organised, connected and stored in long-term memory. Personal accounts of learning are more fluid, describing the combining and recombining ideas along with language, representations, memories of sights, sounds and the learning environment. In mathematics, there are accounts of moments of unexpected insight when no conscious thinking is going on and moments of realisation where new connections and relationships fall into place (e.g. Hadamard 1954). The social processes of being educated and responding to teaching also act to shape knowledge and support the development of critical knowing.

The fact that scientists can now measure activity in the prefrontal cortex neither alters nor explains mathematical thinking. But it does tell us, usefully, that executive function develops throughout childhood and adolescence. Paul Burgess' research group recognises that neuroscience cannot take into account authentic classroom settings over time, so such work is limited (Burgess et al. 2006). Nevertheless, the team usefully contributes some understanding of the development of brain capabilities (e.g. Dumontheil 2014). What emerges is a generalisation that acts as a useful guide for expectations of learners at as they go through school. This is a guide to common features, not a recipe for teaching, nor a description of fixed abilities. In their synthesis, the development of brain capability when people elaborate and evaluate abstract ideas is similar in a range of activities that would be expected when someone does mathematics in anything beyond repetitive routines. Retrieval of episodic memories, manipulation of abstract ideas, maintenance of task rules, anticipating future events, analogical and relational reasoning, reality monitoring and mind wandering are thought to all develop in roughly parallel ways.

According to the synthesis in Table 4.1, infants respond to familiar stimuli in both innate and learnt ways. Pre-school children have usually begun to think about past and future, can deal with situations that have multiple aspects and choose best actions, sometimes with help. Then throughout

Table 4.1 Typical changes in learning states throughout school years (synthesised from Dumontheil 2014)

	Change during school years
Self-generated thoughts	From being able to handle one abstract idea, integrate results of lower order thinking eventually being able to handle several abstract (non-sensory) ideas simultaneously
Response to perceptual stimuli	Can be distracted by perceptual features; becoming less so as attention control strengthens. Know if they have seen something before or not
Thinking with self-generated objects	Improving planning, reasoning and manipulating self-generated objects
Analogical reasoning	From comparing objects to comparing relations
Memorised facts and methods	Becoming able to retrieve alternatives from memory and evaluate their usefulness
Relational and logical reasoning	Improvement consistent by age apart from dip in early adolescence

childhood these capabilities develop and expand. From about 7 or 8 children can generally deal with multiple dimensions of a situation and reason strategically—possibly ignoring irrelevant aspects in the environment. Throughout childhood also the tendency to apply something from one situation to another situation increases. In adolescence, the speed of these processes appears to increase, and more aspects can be dealt with at a time. While even young children can draw analogies and use solution methods in new situations, it seems that their ability often depends on the similarity of objects, whereas older students are better able to see underlying structural similarity that is relational rather than visual. Dumontheil (2014) comments that this is not only age related but also knowledge related, and adults typically use both kinds of reasoning. This research is all based on the Piagetian assumption that brain development leads cognitive capability, where followers of the Vygotskian tradition would claim that cognitive capability, when mediated through tasks, talk and tools, leads development. When I have presented the table to audiences there are always audience-generated examples of incidents in which learners have far exceeded the developmental descriptors. All activity can be enhanced by talk, context and tasks.

Similar changes of knowledge perception occur when adults go through an educational pathway, thus showing that progression in thinking is probably as much due to the novelty of a new subject discipline as age. William Perry (1970) constructed a description of such changes in adults which fits, to some extent, developmental changes in knowing. They might start with a dualistic view of knowledge, whether something is right or wrong, whether an object is an example of a class, or not. To gain this kind of knowledge we are largely dependent on others to tell us somehow, or we construct these duals for ourselves from what others say. 'You can't be three and a half because I am three and a half!' exclaimed a small child who was locked into an inappropriate (in this case) dualistic view of the world. It follows immediately that if mathematics is taught as if the teacher is the judge of all that is true and right, students at any level are locked into a child-world of dualism, leading to such problems as 'is a cylinder a prism?'—a debatable issue. After that, Perry suggests, comes a state of multiplicity for the intellect when it is recognised that there are many possibilities but only one truth from one authority—we might look for differing definitions of 'prism' and 'cylinder'. In the feminist work of Mary Belenky et al. (1986), this can be experienced as a crushing of the emergence of one's own thoughts and ideas. Many people want a cylinder not to be a prism as it has special qualities and therefore deserves a special name; others want it to be a prism because it has many of the same properties. It depends what definitions you choose for both whether it is or is not. Here is one of the problems of mathematics teaching and learning: usually in school there *is* only one mathematically consistent truth so why should students have to construct it for themselves? Take the question of whether '1' is a prime or not. In a world of 'prime' versus 'multiple' it has no role so cannot be a building block for other numbers. I know one young student who claimed that 'when multiplying, 1 is a waste of a number' but had thought, dualistically, that it must be prime. The truth of facts and theorems depends on the internal consistency of mathematics, and hence varied possibilities can be explored legitimately and conjectures developed, modified by conventions. There are some truths of mathematics that have to be taught by someone who knows them, such as the structure, conventions and syntax of symbol systems, knowing what mathematics as a field has found it useful to define, delimit and name. David Hewitt labels 'arbitrary' knowledge as those things that could have been differently expressed, but have become conventional truths. The alternative is 'necessary knowledge' that can be deduced by reasoning using the arbitrary givens; these also have the status of truth but mathematical consistency is the authority. Larry Copes (1982) interprets William Perry's 'multiplicity' state as a learning position in which deciding between several possible pathways could be personal preference, or situational expediency and, for Mary Belenky and her colleagues, this can be accompanied by a growing sense of personal empowerment through informed choice.

Finally, William Perry claims relativism to be an advanced intellectual form in which the student understands the effects of context, the domain of truths and the need for evidence. This rescues his progression from the problem of mathematical consistency—'but mathematics is about given truths'—because this meaning of relativism depends on mathematical experience, context and reasoning which are components of the subject matter, not merely components of pedagogy. For example, in elementary geometry theorems tend to depend on other theorems, so the order in which you prove things affects what you can prove next and how you can prove it. Decisions about whether and when to include the endpoints of intervals depend on the context that has triggered the need for a decision. Knowing the most efficient method of solving an equation, rather than using a universal but possible unwieldy method, depends on experience. Choosing whether and how to substitute a value for π or g might depend on experience, context or anticipations about the next stage of the work. The social world of the classroom, or other learning space, matters not only because it is a source of knowledge but also, for school-age students, a place to work out how to be a thinking person. Perry and Belenky add an extra perspective on learning new knowledge, the transition from total dependence on an authority to personal power.

To me it is obvious that pedagogy and classroom dynamics can take into account the shifts students have to make over time, perhaps over several years, between dualism and relativism. In mathematics, relativism might only be a way of dealing kindly with each other's ideas even if you think they are wrong, and even if they end up being shown to be wrong. On the other hand, it might mean identifying a need for more precision,

Proof 1:

Given: the base angles of isosceles triangles are equal; AC is the diameter of a circle with centre O; angles on a straight-line sum to 180°; interior angles of triangles also sum to 180°

To prove: the angle subtended by the diameter at the circumference of a circle is a right angle.

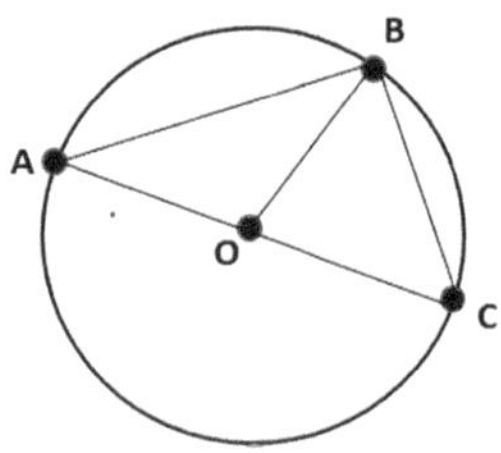

Proof:
Δs ABO and CBO are both isosceles, because their sides are equal (radii of the circle with centre O);
Angle OAB = angle OBA (base angles of an isosceles Δ);
Angle OCB = angle OBC (base angles of an isosceles Δ);
In ΔABO, angle AOB is (180° –2OBA)(angle sum of Δ);
In ΔCBO, angle COB is (180°-2OBC)(angle sum of triangle):
(180°-2OBA)+(180°-2OBC)= 180° (because angles AOB + COB make a straight-line);
Rearranging gives 2(OBA+OBC) =180°. Therefore angles OBA+ OBC= 90°.
This argument does not depend in the position of B.

Proof 2:

Given: Circle centre O with diameter AC and point B on its circumference; the unique diagonal bisection property of rectangles.

To prove: Angle ABC is a right angle.

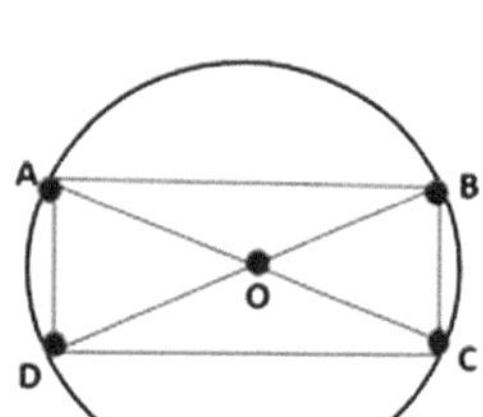

Construction: Draw line segment BO and exend it to meet the circle at D.

Proof: BD is a diameter of the circle. Therefore BD and AC are equal in length (diameters of the circle) and the point O is the midpoint of each. Having diagonals equal and bisecting each other is a unique property of rectangles, so angle ABC is a right angle.

This proof does not depend on the position of B.

Fig. 4.1 Two different proofs for the 'same' geometry theorem

or more definition (as Imre Lakatos points out, 1976) or the understanding that there are choices that might give different solution paths, or different insights while pursuing the same answer, fact or theorem. As an example, I give two proofs of a secondary-school theorem (see Fig. 4.1).

Each proof uses different initial assumptions, prior knowledge and methods, so they do not match each other thought for thought, nor insight for insight. They depend on different perspectives, perceptions and assumptions. Understanding how these operate in mathematics is part of learning mathematics and 'pins down' a relativist orientation towards the subject.

School Ethos

As well as brain function and language, many other factors affect learning: physical, psychological, cognitive, social and emotional factors (Rose et al. 2013) and it cannot be left to a school's general ethos to take care of these somehow. It is possible for a condition described as 'maths anxiety' to develop even in schools where the student is generally happy; for example, it even exists among pre-service teachers (Harper and Daane 1998). Furthermore, a student might be trying to work really hard but for some reason is not focused on what the teacher is intending—maybe they have noticed something else, or lost concentration for a critical moment, or they know too much or too little to engage with the idea. Maybe the idea is being presented only verbally and there are barriers to understanding or maintaining attention through listening. 'All eyes facing front' does not imply attention to an idea, nor do teachers who 'drone on and on' always enable attention.

What seems to help manage negative emotions while doing mathematics is to move away from immediate situations that trigger negative emotion and review them from safer ground, for example, being given an answer to check or explain, or approaching from different direction, or having control of a self-checking mechanism, or having some choice. The problem then is how a student will make necessary advances if they are the ones choosing without an overview of what might be made possible, or prevented, by their choices. It cannot be left to chance whether everyone has the opportunity to learn particular ideas. Overlaid onto this are all the other aspects of school: liking, or not, the teacher and the subject; feelings of competence; willingness (or not) to engage with test-focused or self-sustained pathways; prioritising other things over school, or this lesson; hunger; comfort in the presence of other students.

Adolescence sits in a continuum of development and combines changes in young people's place in society and physical changes concurrently with changes in the brain while more abstraction and complex thought becomes possible. In many ways, adolescents are capable of functioning independently. Physically they are capable of everything needed to stay alive, to earn money, to feed and clothe themselves and make decisions about what to do and how to do it. Physically they become able to create children. Legally they are minors and have no democratic power and few entitlements. In the midst of this, education is both a right and an enforcement.

While psychological research about young people is usually based on task-related observations, research on adolescents is often based on self-report, interviews, observations of behaviour all of which are influenced by the prevailing culture and times within which the research is carried out. For example, as I write answers to questions about hopes and fears might centre on climate change whereas in the 60s they might have centred on nuclear war. Answers about imagined future employment might vary according to current economic circumstances; answers about adult life might reflect current societal and cultural norms. The Freudian psychologist Erik Erikson (1968) suggests that adolescence is a time when the development of personal identity involves: sometimes adopting extremes of dress, behaviour, social choices; forming stereotyped relationships; avoidance of planning for the future; working obsessively or not enough. Using this information reliably requires questioning whether these traits are specific to the 60s and/or to adolescents. Critics have pointed out that conclusions about development are predominantly the insights of male researchers and influenced by European Heritage culture; development might be different for women and in other cultures. Nevertheless, Erikson provides a starting point. He characterised several psychological states of development to adulthood, of which three pertain to school years. Engaging in purposeful activity while sometimes launching out independently and getting things wrong is seen as a feature of early years, becoming competent and aware of competence, and hence also of incompetence, is a feature of primary years; making commitments to people and groups, but possibly getting confused about where to place loyalty, and developing selfhood are features of adolescence. These descriptions depend on pathologising typical psychological conflicts in each state, as if conflict is a 'norm' either as a driver or outcome of each developmental stage. This may sound unduly negative but, as Sarah Jayne Blakemore points out (2018):

> Three-quarters of mental illnesses appear before the age of 24. Depression, anxiety, eating disorders, schizophrenia. ... Changes are happening at the same time, a confluence, a perfect storm: hormonal changes, neural changes, social changes and the pressures of life suddenly increase. The changes in the brain mean it is particularly plastic and susceptible to environmental stress. The pressure of schools and exams is increasing.

John Coleman (2011a), a clinical psychologist with extensive hands-on experience of working with troubled teenagers in the UK context, claims that identity is generally more stable during adolescence than these models might imply. The fact that stability is found among adolescents who have had a hard time seems encouraging. He developed a 'focal model' of adolescence in which 'coping' involved 'focussing on different issues at different times, spacing out the problems and concerns they had to face.' (Coleman 2011b: p. 17 ff). His work has practical implications for teaching adolescents, in that they can focus on something that is sufficiently interesting in competition with other, potentially difficult, life factors. Capacity to inhibit irrelevant factors from a current task becomes stronger throughout childhood and the use of the brain's natural executive powers is subject to agency and will (which could also lead to compliance or rejection).

What does this mean for teachers of mainstream classes in which a one-size-fits-all model of pedagogy is normal? While Coleman was learning from troubled teenagers he observed, over time, that the nurturing commitment of adults mattered to the healthy development of agency and will (Coleman 1980). That this positive view of development arises from observing people in trouble suggests that it can also be transferred to those in more stable and caring contexts. He observes that disenchantment with the education system often emerges during secondary school when there is little opportunity to be in control of anything. Increasingly, adolescents question the purpose of education if they cannot see how it relates to their future and fails to give an awareness of learning. Difficulties in relationships with teachers can arise because in secondary school they typically have many different teachers rather than one central teacher as is usual in primary schools. Furthermore, having been the oldest in primary school they may feel they are not given enough responsibility in secondary. The reaction of their peer group and parents to these changes matters and can support a positive sense of growing up. Adolescents, therefore, need not be passive recipients of schooling. They can critically evaluate and use school-based opportunities.

Coleman summarises the connections between development and learning in secondary school as need for 'gradually increasing responsibility and autonomy, the need for more equal relationships between adults and young people, the need for some element of individual choice in relation to academic learning, the need for rules and activities to have a clearly perceived rationale, and the need for opportunities that offer space for the young person to demonstrate maturity' (2011b, p. 173). In future chapters, I shall report some teaching that has these ingredients.

I see important links between adolescent psychology, the content and expectations of typical school mathematics, and the development of brain function described by Dumontheil (2014). I have described sense making in mathematics as a discipline as changing from:

- understanding what is familiar to what is unfamiliar, possibly with new tools, classifications and ways of seeing;
- observable and tangible features to imagined, abstract features, introduced through language and tasks;
- intuitive responses to reasoned responses;
- immediate to mediated reactions;
- focusing on objects to focusing on relations between objects;
- understanding relations among sensory objects to relations among imaginary objects.

All of these become more necessary as the curriculum progresses and offer growth of abstract personal power when this is not only the desire of students but also a developmental possibility and drive (Watson 2010). To learn mathematics we have to change the ways we make sense: to handle unfamiliar abstract situations as well as familiar ones; to reason with data beyond that which is sensory; to know when to allow reason to trump intuition; to think about social and abstract implications; and to act in ways that are socially mediated they can also use these tendencies in school mathematics.

Monique Verhoeven, Astrid Poorthuis and Monique Volman, M. (2019) have reviewed literature about school type and its effects in the development of identity, including subject-specific identity as a mathematician. They found nine studies that were concerned with mainstream mathematics, where the subject matter is 'constrained by tools, norms, and values' (ibid. p.44). One of these involved analysis of data from a study that compared the practices and effects of different kinds of mathematics teaching over a period of five years in the USA (Horn 2008). Broadly speaking, the two kinds were collaborative groupwork in classes that were not segregated by prior attainment and individual work on short problems designed to provide a cumulative, linear, development of mathematical competence. In the second type of school, one teacher explained to a low-attaining class that:

> Math is something that you only have to remember everything that you've ever learned before. And you get to a point somewhere along the line where your brain says, 'My brain is full.' And you can't go on. (Horn 2008, p. 220)

In the first type of school, the attitude was exemplified by

> You know, there's a sense like: 'All of you can accomplish this.' You know, it's not just a few. And I think the fact that it isn't tracked kind of makes that real clear to students. [. . .] everyone just assumes that students should and can achieve, and can do the work. (ibid. p. 222)

In the first case, a more flexible view of the capacity to do mathematics, and the nature of mathematics, is held alongside an exploratory approach. Emphasis is placed on big ideas in mathematics and the methods of exploration and inquiry that give everyone access to them. Ilana

Horn's research is reminiscent of a similar study undertaken in England in the early 90s (e.g. Boaler 1997) in which similar differences were also examined. Horn's focus on particular students whose attainment trajectory in mathematics changes from 'low attainment' to 'college ready' during the study is very valuable, because she analyses their perception of the subject, of themselves as students of mathematics, and shows how these relate to their sense of self-worth and competence more generally. Even in this brief summary of her study different ways of knowing mathematics and the teachers' differing epistemological understanding are shown to play a role in the students' perception of their own capabilities and their eventual attainment. However, this finding cannot be reduced to 'group exploration good; direct individual instvruction bad' since these both have a wide range of enactment and also address different aspects of the nature of mathematics and mathematical activity. But Verhoeven et al. do conclude that 'Providing explorative learning experiences is [...] a pedagogical approach that stimulates adolescents to connect what they are taught in school to who they are and want to be.' (2019, p.59).

Reflections on Content and Pedagogy

At the end of the previous chapter, I proposed that, in order to develop independent learning and the ability to deal mathematically with unfamiliar situations, learning mathematics needs to include, alongside knowledge of facts and methods:

- knowing how and when to act mathematically;
- having several ways to access, use and restructure memory in mathematically relevant ways;
- knowing how to work with unfamiliar mathematical situations;
- knowing how to use mathematical methods and tools to manage cognitive load;
- knowing how to coordinate knowledge gained from everyday experience with formal, symbolic approaches to mathematics.

In this chapter, generic pedagogic approaches have been described that take these into account in various ways, some prioritising upskilling over reasoning for social justice purposes, some developing critical thinking tools for similar reasons or for general mathematical benefit. I have attempted to break down unhelpful dichotomies that arise because research is easier in simplified and controlled pedagogical contexts than in complex pedagogies involving fluid interactions. I situated school mathematics within the psychological, emotional and social worlds of school-age students, pointing out how mathematics relates to, and might enhance, development and how the school context matters.

The idea that cognition can be controlled seems to lurk under the surface of some applications of research, but it is beginning to look as if 'care' might be a more appropriate approach. In the next chapter, I will explore this idea.

References

Abadzi, H. (2006). *Efficient learning for the poor: Insights from the frontier of cognitive neuroscience*. Washington, DC: World Bank.

Belenky, M., Clinchy, B., Goldberger, N., & Tarule, J. (1986). *Women's ways of knowing*. New York: Basic Books.

Blakemore, S. (2018). www.theguardian.com/science/2018/mar/25/sarah-jayne-blakemore-secret-life-teenage-brain. Accessed 11 Apr 2020.

Boaler, J. (1997). *Experiencing school mathematics: Teaching styles, sex and setting*. Buckingham: Open University Press.

Burgess, P., Alderman, N., Forbes, C., Costello, A., Coates, L., Dawson, D., Anderson, N., Gilbert, S., Dumontheil, I., & Channan, S. (2006). The case for the development and use of 'ecologically valid' measures of executive function in experimental and clinical neuropsychology. *Journal of the International Neuropsychological Society, 12*, 194–209.

Bruner, J. (1960). *The process of education*. Cambridge, Mass: Harvard University Press.

Chen, O., Kalyuga, S., & Sweller, J. (2015). The worked example effect, the generation effect, and element interactivity. *Journal of Educational Psychology, 107*(3), 689–704.

Coleman, J. (1980). Central issues in the residential care of young people. *Journal of Adolescence, 3*, 175–186.
Coleman, J. (2011a). *Why adolescence matters*. http://www.jcoleman.co.uk/why-adolescence-matters. Accessed 11 Apr 2020.
Coleman, J. (2011b). *The nature of adolescence*. Abingdon: Taylor & Francis.
Copes, L. (1982). The Perry development scheme: A metaphor for learning and teaching mathematics. *For the Learning of Mathematics, 3*(1), 38–44.
Dienes, Z. (1969). *Building up mathematics*. London: Hutchinson Educational.
Dumontheil, I. (2014). Development of abstract thinking during childhood and adolescence: The role of rostrolateral prefrontal cortex. *Developmental Cognitive Neuroscience, 10*, 57–76.
Engelmann, S. (posted 2013). https://www.youtube.com/watch?v=j9SjFsimywA. Accessed 11 Apr 2020.
Engelmann, S., & Carnine, D. (1982). *Theory of instruction: Principles and applications*. New York: Irvington Publishers.
Erikson, E. (1968). *Identity: Youth and crisis*. New York: WW Norton.
Gattegno, C. (1961). https://www.youtube.com/watch?v=HMetsXgMd5o&feature=emb_rel_end. Accessed 11 Apr 2020.
Gramsci, A., & Hoare, Q. (1971). *Selections from the prison notebooks*. London: Lawrence & Wishart.
Hadamard, J. (1954). *The mathematician's mind: The psychology of invention in the mathematical field*. Princeton: Princeton University Press.
Harper, N., & Daane, C. J. (1998). Causes and reduction of math anxiety in preservice elementary teachers. *Action in Teacher Education, 19*(4), 29–38.
Hodgen, J., Foster, C., Coe, R., Brown, M., Higgins, S., & Kuchemann, D. (2020). *Low attainment in mathematics: an investigation of Year 9 students in England*. https://www.ucl.ac.uk/ioe/departments-and-centres/departments/curriculum-pedagogy-and-assessment/low-attainment-mathematics-investigation-year-9-studentsengland. Accessed 8 July 2020.
Horn, J. (2008). Turnaround students in high school mathematics: Constructing identities of competence through mathematical worlds. *Mathematical Thinking and Learning, 10*(3), 201–239.
Kaufman, M. (1973). *The effect of the DISTAR instructional system: An evaluation of the 1972–1973 Title I program*. //eric.ed.gov/?id=ED110170. Accessed 11 Apr 2020.
Kim, Y. (1972). *DISTAR instructional system*. https://files.eric.ed.gov/fulltext/ED061632.pdf. Accessed 11 Apr 2020.

Krutetskii, V. (1969). An investigation of mathematical abilities in schoolchildren. *Soviet Studies in the Psychology of Learning and Teaching Mathematics, 2*, 5–57.

Kullberg, A., Runesson, U., & Mårtensson, P. (2014). Different possibilities to learn from the same task. *PNA, 8*(4), 139–150.

Lazonder, A., & Harmsen, R. (2016). Meta-analysis of inquiry-based learning: Effects of guidance. *Review of Educational Research, 86*(3), 681–718.

Lester, F. (1994). Musings about mathematical problem-solving research: 1970–1994. *Journal for Research in Mathematics Education, 25*(6), 660–675.

Marton, F. (2014). *Necessary conditions of learning*. London: Routledge.

Noddings, N. (1993). Politicizing the mathematics curriculum. In S. Restivo, J. P. van Bendegem, R. Fischer (Eds.), *Math worlds: philosophical and social studies of mathematics education* (pp. 150–161). Albany, NY: Suny press.

Paas, F., & Sweller, J. (2011). An evolutionary upgrade of cognitive load theory: Using the human motor system and collaboration to support the learning of complex cognitive tasks. *Educational Psychology Review, 24*, 27–45.

Perry, W. (1970). *Forms of intellectual and ethical development in the college years: A scheme*. New York: Holt, Rinehart, and Winston.

Piaget, J. (1936). *Origins of intelligence in the child*. London: Routledge & Kegan Paul.

Rose, L., Rouhani, P., & Fischer, K. (2013). The science of the individual. *Mind, Brain, and Education, 7*(3), 152–158.

Rosenshine, B. (2012). Principles of instruction: Research-based strategies that all teachers should know. *American Educator, 36*(1), 12.

Verhoeven, M., Poorthuis, A., & Volman, M. (2019). The role of school in adolescents' identity development: A literature review. *Educational Psychology Review, 31*(1), 35–63.

Watson, A. (2010). Shifts of mathematical thinking in adolescence. *Research in Mathematics Education, 12*(2), 133–148.

5

Care

Need for Care

In 2019, Sir Tim Brighouse, eminent UK educationist who had been involved in improving education for all young people for decades, and had been honoured for it, wrote that he imagined it would be possible to have 'a curriculum designed to develop values, attitudes and ideas as well as knowledge and skills … in short, affecting their hearts and hands as much as their minds'.

Was this a call to have a vocational, practical approach to education? No, because he also mentions hearts. Is it then merely a romantic view of learning, education and childhood that has been attacked frequently by critics of Rousseau and Dewey? No, because cognitive scientists also accept that emotion and physical engagement play a part in the effort of learning and can influence the extent of what is possible in working memory, and also in schema-formation (e.g. Damasio 1999). This does not only mean whether you like mathematics, enjoy lessons and appreciate your teacher—it also means that some emotional mechanisms are always at work when we are given, or take on, a new task. Neuroscience, once it gets its subjects' heads out of the scanners, also supports the view

A. Watson, *Care in Mathematics Education*, Palgrave Studies in Alternative Education,
https://doi.org/10.1007/978-3-030-64114-6_5

that emotion is a component of learning (e.g. Roebers 2017). So hearts and hands and minds are all necessary for learning. Is the whole of young people's development therefore an appropriate goal for education systems? No one seems to deny this, and school mission statements frequently mention such goals. I obtain the following by dipping into the Internet: 'develop the empathy and drive to make the world better'; 'develop responsible, determined individuals who love to learn'; 'become the best version of ourselves'; 'become capable, caring and confident'.

My focus is to place mathematics subject knowledge firmly into the need to care for whole, healthy and self-deterministic development. As a rationalist and a mathematician I find stories about student-centred pedagogy in the creative, practical and liberal arts neither inspiring nor relevant when thinking about learning mathematics. I want to understand the roles of hearts and hands and care in learning the abstract structured 'given' discipline of mathematics. Are students coerced into working effortfully in a rule-based context to pass examinations, or to please teachers, or to conform to institutions and hence keep out of trouble and get smiles and certificates? Such coercion could be seen as care for learning when learning is evidenced by test results, but what about the development of brain function and properties of mind that can engage powerfully with the world through mathematics? Can it confuse students if their role models for mathematics, their teachers, make out that something is easy when in the students' experience it has no meaning, is hard to remember, and even a source of upset and shame?

About a decade ago it occurred to me that trying to make students think in particular ways rather than develop their own ways of thinking might be a form of low-level bullying, ignoring or negating the student's own thinking. I called this 'cognitive bullying' (Watson 2011) and suggested that it could cause stress, undermine efforts to learn, lead to repeated failure, rejection of the subject, and anxiety or, at best, mute compliance to rote learning. Definitions of bullying vary but in general agree that it involves coercion through repeated and habitual domination in unequal power relationships, threat, abuse, and leads to fear, anxiety and loss of self-esteem, often intentional. My use of the term does not include intention, but the term is also used to explain the feeling of having been bullied, even when there was no intention on the part of the perpetrators. Some other writers have used the term 'cognitive abuse' (Johnston-Wilder et al. 2010). Ruth Merttens claims that any teaching that focuses solely on the student rather than shared experience of mathematics can be interpreted as a form of harassment (private communication, Feb 24, 2020). These phrases will strike dissonant chords with some readers. I wanted to draw attention to the effects of teaching that does not take students' thinking into account. It concerns students who begin to exhibit symptoms of avoidance, resignation, non-compliance, rejection, flight, fear and anxiety at some stage of their mathematics schooling. In extreme cases of 'maths anxiety' heartrates rise, breathing is difficult and palms sweat when mathematical activity is expected (Ashcraft 2002). The release of brain chemicals associated with fear inhibits learning. Socially, many adults cope by seeking resonance among others and using humour or solidarity about their rejection of the subject. How do students reach this state?

In subsequent discussions about this, it occurred to me that talking about cognitive coercion only made sense if I could also describe cognitive care. Of course teachers care for their students, but as a frequent observer in a wide range of classrooms I often see overtly caring environments being constructed with the aim of persuading students to act only in prescribed ways in relation to mathematics, rather than environments in which the perceptions and thoughts of students are heard and respected and used as the clay with which robust mathematical thinking could be moulded. Brent Davis chronicles a teacher, whom he calls Wendy, through vignettes showing changes in her practice. When he first observes a lesson she follows her version of a script—a detailed lesson plan she has designed saying, 'I'm quite proud of this lesson. I've almost got it to the point where it feels like everything's in place' (Davis 1997 p. 361). The individual students and their perspectives play no role in the unfolding of the lesson; she checks with them every now and then to ensure they are listening to her. Brent Davis calls this 'evaluative listening' as Wendy evaluates their acceptance of her message. Through encouragement to think about this she changed her ways of questioning to an information-seeking mode and listens more to the way they articulate mathematics, asking for elaboration and explanation. What they say does not alter her planned direction of the lesson however as she is listening for particular responses that enable the lesson to proceed. She is paying attention to their sense-making so that she can avoid ambiguities and complications on the way to her desirable learning goal. Davis calls this 'interpretive listening'. In a third vignette Davis reports that

> The rigid structure, the preset goals, the formulated explanations, the entrenched roles have all been dissolved to some extent as the unanticipated, the surprising, and the intuitive find places for expression. The mode of teaching seems to be more a matter of flexible response to ever-changing circumstances than of unyielding progress toward imposed goals. The importance of listening in enabling the teacher to respond is evident (ibid. p.369).

Her listening had become 'hermeneutic'; Wendy was now responding to students rather than the only responses being from the student to the teacher. It is worth remembering that some students are not listened to seriously at home or other places in addition to school classes.

As I write this I can imagine objections being raised—for example, in team sports it would be counterproductive to listen to everyone's perceptions and thoughts yet most students seem happy to participate. PE teaching includes recognising the strengths and weaknesses of individuals and coordinates these for the benefit of the team while helping individuals to develop. There is also a clear shared task. There are others for whom team game lessons are miserable experiences but who might flourish with activities that allow a more individual focus such as track sports. With mathematics, however, there is no alternative to learning the conventions and meanings of the number system and algebraic notation—without these they are not learning mathematics. For full engagement in everyday life people need numbers and quantities, and some of the arithmetic and representations that go with these. Anxiety about learning these emanates not only from the students themselves but also from parents, carers and teachers and is institutionalised by the state in its testing regimes. What sort of care can alleviate such anxiety, or prevent its development?

Nature of Care in Education

For Martin Heidegger (1927), writing about how human beings are grounded in the realities of the world, the fundamental basis of humanity is care. None of us are neutral. In education, non-neutrality has many forms, can be about teaching or learning, and is generated in relations between human beings who care about each other to some extent. Nel Noddings, who was a secondary school mathematics teacher before becoming an academic philosopher and a prolific writer on care, says: 'When I care, I really hear, see or feel what the other tries to convey' (Noddings 1992, p. 16). In her view, care requires one person to pay total attention to another, even momentarily, parking their own motivations to become 'engrossed' in the other and their needs. In thinking about education, she expands this meaning to include helping others to care, specifically helping students to become caring people in general and also to care about their own learning. Noddings believes that a person's capacity to care can, in some respects, vary according to the object on which they are focusing, and also according to prior knowledge and experience. Because caring is a matter of relationships, the capacity to care about particular people, ideas or things can develop through expression of people's interests and enthusiasms. If I communicate my enthusiasm for mathematics within a positive and caring relationship, those for whom I am caring can understand that such enthusiasm is possible and may even join in. Communication between teachers and students is a conduit for care, and also for developing the capacity to care. Of course communication can also be a conduit for reducing or warping what appears to be care into something quite negative and even damaging; fairy tales are full of these tropes. This is why thinking about teaching as a caring profession is so important, not only for nurturing intellectual subject-based development but also for the wellbeing of students.

Noddings develops a layered model of foci for care (1992)—a kind of onion that has at its heart care for oneself and works its way through existence to a final layer of caring for ideas:

- Caring for self;
- Caring for the inner circle;
- Caring for strangers and distant others;
- Caring for animals, plants, and the earth;
- Caring for the human-made world;
- Caring for ideas.

Caring for oneself is a major focus for teaching young children who need to understand how to survive, stay healthy and fit, value themselves, appreciate learning and know how to act in the world. Caring for the inner circle of people with whom they have regular communication fits with models of social and emotional development for young people into adolescence. Neither of these layers vanish; care extends to include unknown others and also all life and the environment. She also wants students to care about what can be made by humans—everything from arts and crafts to robots and vaccines—and believes that such care arises from actually making things and being alongside others who make things. Finally, she wants students to care for ideas.

She is writing in a US educational context at a time when the agreed way of educating everyone is to provide a curriculum focused on verbal and mathematical ways of knowing, with arts, humanities, social and practical subjects being undervalued in comparison. Against this, her primary aim is to develop competent and caring people rather than only people who all have some kind of shared factual knowledge (ibid., p. 154). Thus, becoming caring in a broad sense is prioritised over abstract ideas, and the ideas about which she is most concerned in this model are values and virtues—not school subject ideas per se.

A Relationship Between Learning and Care

Given that she was a teacher of mathematics I find her position on mathematics difficult. She argues that students only need the mathematics that relates to their interests and possible futures. This argument has been made for decades and has often been connected to students' supposed limitations of intellectual capacity. In England, the education system used to be based on discriminating at age 11 between those for whom an abstract intellectual education was appropriate and those for whom a more practical and vocational education might be more appropriate. The process of separating students with different test results still exists in the independent school sector, and in some local jurisdictions, and is currently increasing in state-funded schools as selectivity creeps back under various guises (Walford 2014). For mathematics, this process has never really gone away since even schools espousing comprehensive principles set students into different teaching groups for mathematics according to prior test results. Ostensibly this is to separate students into groups that may need more challenge or more help than others, but in practice these usually become groups with different learning expectations and aspirations (Wiliam and Bartholomew 2004). Noddings, however, is not making the argument on the basis of innate capacity and past test results but on an argument about care that requires teachers to listen attentively to students and work out how to respond to their interests and intentions. In particular, she argues in various places that not everyone is going to need algebra, geometry and trigonometry in their lives, whereas everyone needs to understand percentages, ratios, financial and statistical arguments (e.g. 1992, P.172). Students, she says, should be allowed to make choices about whether to study mathematics beyond her list of useful topics or not because understanding any subject grows from interests and purposes. In most schools there is choice in subject pathways from mid-adolescence onwards, but mathematics has nearly always been compulsory except in some free-thinking independent schools. Choosing one's own pathway is certainly more empowering and democratic than being forced to do mathematics because it is good for the economy, but this can be a move that limits some people's mathematical progress if it is only informed by poor relations with teachers, low past achievement, a limited view of the subject so far, and a limited view of possible futures. There has to be provision of flexible later learning opportunities to justify a model of personal choice, but for most people the provision of mathematics teaching, like provision of reading tuition, is front-loaded into childhood and not available on a 'when I need it' basis.

I agree with her that what is unhelpful is equal but superficial access to the knowledge that people who are already privileged regard as their own in some way. In England, this would relate to nineteenth-century novels in English, opera, colonial history from an English perspective, and so on. Of course such superficiality would be limiting for everyone, but, Noddings argues, is inevitable if you expect everyone to engage with ideas in which they are not interested and for which they cannot see the connection to their own current and future lives. Thus, she opposes the notion that a full mathematics education including trigonometry, algebra and calculus is necessary for the development of shared cultural capital across social classes—a current aim in England as I write—because of the impossibility of successfully teaching everyone a deep understanding of the power of those aspects of mathematics.

It is not clear to me how anyone can choose what they need to learn for the future if they are not aware of the fundamental meanings of, and in, mathematics. In some primary schools, children learn little more than arithmetic, especially calculation, and the naming of geometrical shapes and their features. In addition, they learn something about using data to make decisions, and using calculations to answer questions about quantities. They may do a little logical reasoning, maybe about angles or the number system, and some pre-algebraic work in which some quantities are not known. In short, they have only just begun to experience the real business of mathematics which is to model relationships between quantities in the real world, and to appreciate abstract structures in the mathematical world, and use these to create more knowledge through reasoning. How can they know whether they are ever or never going to find these useful or interesting? In practice, it is education systems, in the form of teachers, selection methods and testing, that decide on someone's needs rather than they themselves, whether ignorant of the implications or not (Boaler et al. 2000). Anticipation of the future *shapes* the mathematical future of the student and the rapid rise and complexity of digital capabilities has brought within everyone's reach the mathematical tools that used to be the preserve of the few. One particular change has been access to microworlds through games, with their own rules, syntax and semantics, so that most learners have had some experience of entering a symbolic world and making it work for them—similar to entering the worlds of formal geometry or algebra. Before this, structured imaginary non-mathematical worlds were only available through reading and board games; now they are accessible and accessed by most. This gives lie to the belief that only some learners may be capable of acting competently in the abstract structured world of mathematics. Instead people who work at the edge of how digital worlds can enable mathematical learning look for ways that replicate the personal engagement that, in the games world, can be achieved through avatars and the personal goal-setting that is part of, say, Minecraft ™.

In my view, choice in school mathematics only makes sense if students' experience of mathematics so far has been of an array of authentic mathematical tools used in a variety of ways, including those abstract ways that make the modern world. Noddings says that caring relations can prepare students for a range of experiences, through the establishment of trust. Educators can influence interest for mathematics, and in my view they should. This is why care for students and care for the subject matter need to be intertwined.

Amy Hackenberg (2005) holds caring and learning together as one system. This idea is built on understanding 'care' to be the recognition of cycles of depletion and the provision of stimuli to fill the emptiness. Depletion can be emotional or cognitive or both. Such cycles are at work when a new idea is learnt; initial senses of puzzlement, emptiness and wonder or worry are resolved when success is felt. Provision of tasks and their successful resolution happens in cycles that are recognised by teacher and student together. The expected outcome is a feeling of satisfaction through learning something new. Care and learning are like two sides of the same coin in these cycles. A lack of resolution of the puzzlement, emptiness and wondering can lead to dissatisfaction unless a classroom culture of suspense has been developed to keep students wondering between lessons. For example, in the DISTAR system students were asked explicitly to do 'thinking at home' and even to report back on their home thinking. Hackenberg's model fits with Jean Piaget's description of learning new ideas—the initial assimilation of new ideas into existing schema that have been constructed from previous ideas and the eventual accommodation of the new by modifying the schema in the light of the new knowledge, leading to integrated conceptualisation. The actions of a teacher to support this process show care for cognition as well as care for the student.

This symbiotic model of care allows for the teacher to care naturally for the student and the mathematics, and also allows for what Noddings calls 'ethical care' (2012) in which teachers may not feel natural personal care but nevertheless continue to act in a caring fashion because it is the professional responsibility to do so. Through ethical caring, natural care might be repaired. Neither Noddings nor Hackenberg underestimate the difficulties in this approach.

The Place of Care in Teaching Mathematics

For me, the job of teaching mathematics requires care for the ideas of mathematics to be much closer to care for the student than Noddings' layers might suggest. Hackenberg offers one way to hold caring and learning tightly together in classroom cycles. Ruth Merttens (personal communication 2019) offers a different way to think about care: that of 'caring with' rather than 'caring for' the student. This view posits mathematics as a shared task in which teacher and student focus together sometimes as equals; sometimes as expert-apprentice. In her image, the shared task is overtly mathematical, the ideas, methods and meanings are in the classroom mathematical culture, the goal is mathematics learning and the actions are those of mutually caring about the mathematics.

For example, 'what can you say about the remainders you get when you divide numbers from 1 to 35 by 7?'. Such a task requires not only knowing at least one method for division, but also typical mathematical methods of inquiry, including discussion with each other or the teacher, the ability and willingness to make effort in a sustained task, and exploration and explanation of emerging patterns. Apart from knowing methods that might be deliberately told, the rest of the requirements can be sustained by what Merttens calls 'passionate attention' (personal communication 2019); Noddings talks about being 'engrossed' in the individual learner (1984, p. 19). The claim is that if passionate attention, that is the kind of attention that is emotionally strong enough to ignore distractions, is focused on individual students instead of mathematics it can lead to precipitate judgement about individual students. For Merttens, the focus of attention is the mathematics. Passionate attention to the task enculturates students into mathematical methods of inquiry, and this attention can become becomes theirs too with teacher and student both focusing in the shared task. Hackenberg's model of symbiotic caring and learning could generate passionate attention on the mathematics. Such attention requires careful listening and observing, not to focus on whether a student is auditing the teacher's mathematics but so that a teacher can understand a student's mathematics and hence engage in two-way communication. This involves sharing the understandings of both the student and the teacher as in Feuerstein's practice of mediation, and Davis' comments about Wendy's ways of listening. To achieve passionate attention I need to care so much about mathematics that I long for my students to understand my care for the subject and also to experience it for themselves through shared work.

What Happens in Educational Spaces

These considerations about care affect how the iconic didactic triangle might be viewed (see Fig. 5.1). This triangle is a common representation for the relationships between teacher, student and content generally attributed to Comenius (1657). Educationists have interpreted,

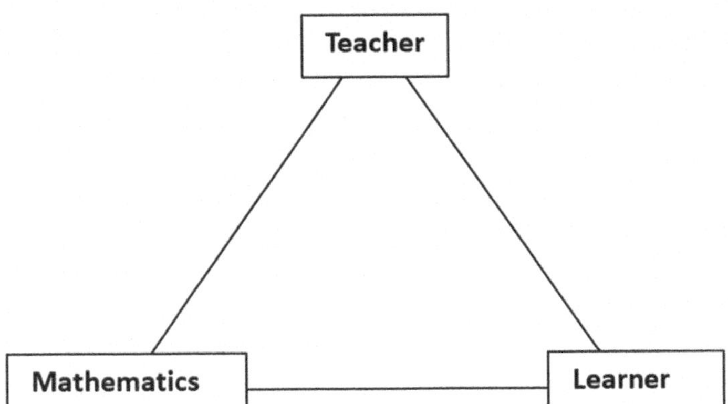

Fig. 5.1 A didactic triangle connecting teacher, learner and mathematics

elaborated, embellished or reconstructed triangular relationships such as this for various purposes.

Guy Brousseau (1997) inserted the idea of a milieu into the mix. 'Milieu' means the physical environment, epistemological norms, and the tools and talk that students accept as how mathematics is to be done and learnt. In other words, those things that shape the ensuing mathematical work and within which the student gets feedback about how their approach fits the teaching intentions. While this is usually set up by the teacher initially, responsibility for how the task is done is handed to students to a greater or lesser extent. 'Milieu' forms a dyad with the students in learning situations, for example a teacher might step back and allow the cultural norms and the inherent logic of mathematics to provide feedback for the student. The students get a feel for whether they are making mathematical sense, using mathematical methods of inquiry, appropriate language and symbolism, and their own reasoning. However, in a situation that requires students to adopt new-to-them conventional symbols or definitions a different milieu might be constructed in which feedback comes from correct use being recognised by the teacher. The word 'milieu' could also be used to encompass the school mathematics culture—the ways in which mathematics is understood as a discipline and hence what teachers and students are expected to do—within which the whole triangle sits. Schoenfeld (2012) elaborated this idea further to emphasise the importance of the backstory of creating the culture that shapes the interactions implied in the triangle and the nature of school mathematics itself. Different layers of culture—nation, community, school, classroom—affect the operations of the basic triangle. The triangle does not itself prescribe specific practices. I want to embed the basic triangle beyond the classroom and into the school and the surrounding communities within which students' lives are embedded.

Within the triangle Noddings, perhaps, might have an arrow pointing towards the student from the teacher, maybe via the subject, while Merttens might have arrows from both teacher and student towards the subject (see Fig. 5.2). Indeed, in her use the teacher might be replaced by parent or carer who takes on the role of teacher or co-explorer when doing mathematical tasks at home.

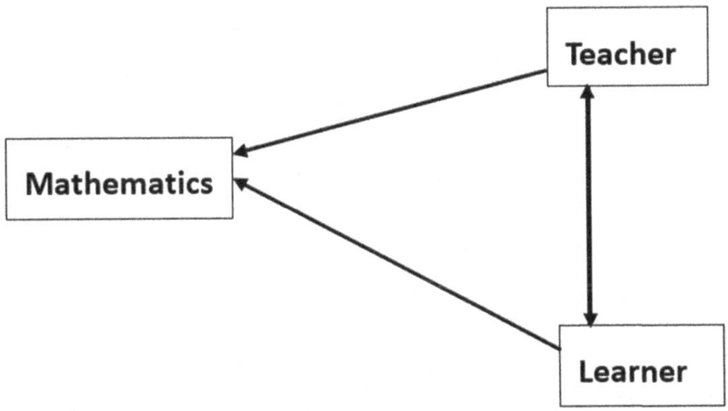

Fig. 5.2 A didactic triangle with teacher and learner alongside focusing on mathematics

Merttens says (private communication 23rd Feb 2020):

> It is tempting to think of the pedagogic relation as the teacher focusing on the child/student. This is an error … because the learning is what happens between people. It is not transferred from one person to another … So the teacher and the student have a joint focus on the object – in this case, the mathematics. Their shared focus is where the learning takes place – between them. (her underlining)

This sense of mutual action within mathematics does not apply to the whole of mathematics learning, as she points out herself. For example, in the task I posed above ('what can you say about the remainders you get when you divide numbers from 1 to 35 by 7?') there has to be at least one known method of dividing, otherwise the exploration cannot be shared. How does a student get to know methods of division? More generally, how does the caring teacher ensure the students has the essential tools for the intellectual task, or maybe a toolbox from which tools can be selected, tested and used? Merttens addresses this by recognising that certain facts, patterns and processes can only acquired by being told or shown with a focus on creating the kind of structured memory for them that ensures they are retrievable when necessary or helpful. In her pedagogy this is often achieved through chanting, story and diagram, her particular kinds of chanting, such as counting backwards in halves, being associated with the underlying mathematical structures being learnt (Merttens 2001). For her, dependency on memory for some mathematical facts is a recognisable feature of mathematics for parents and carers and hence an important cultural link between school and home. Dave Hewitt's distinction, mentioned earlier, between arbitrary knowledge, that is the distinctions, labels, categories that have been created by humans and have to be learnt as conventions, and necessary knowledge, that is what can be deduced logically from working with mathematical structures, is relevant here (Hewitt 1999). So, for example, the number system uses nine digits and zero that are learnt used in a certain order, and then some symbolic combination rules kick in that also have to be learnt so I can count as far as I like. But once I know that 6 follows 5, and the combination rules, I can deduce that 26 follows 25, 1396 follows 1395 and so on—this can be deduced and is necessary for mathematics to be consistent.

Barbara Jaworski (1996) ascribed qualities to three vertices of a Teaching Triad to focus on the work of teaching as she observed it and hence to understand mathematics teaching and to inform professional development: management of learning (ML), broadly the organisation of the classroom environment, tasks and activity; sensitivity to students (SS) being the teacher's knowledge of students that guides interactions; and mathematical challenge (MC) that is offered to students to prompt mathematical activity and thought (see Fig. 5.3).

In her work, 'mathematical challenge' is usually identified as direct engagement with concepts and meanings, often shifting authority to students, rather than procedural imitation, memory and performance. 'Sensitivity to students' is not only about being kind, but about understanding them as students of mathematics. MC and SS seem to be interdependent and she and Despina Potari developed a sense of how these need to be 'in harmony' (Potari and Jaworski 2002) for either to be beneficial in teaching situations that might be described as 'investigational'. It is in the midst of this idea of 'harmony' that I locate 'care for the learning of mathematics' as the overlap between care for students and care for mathematics in Fig. 5.4.

'Care for learning' looks on the students within learning situations and also looks on the mathematics that is being learnt. It is about the teacher

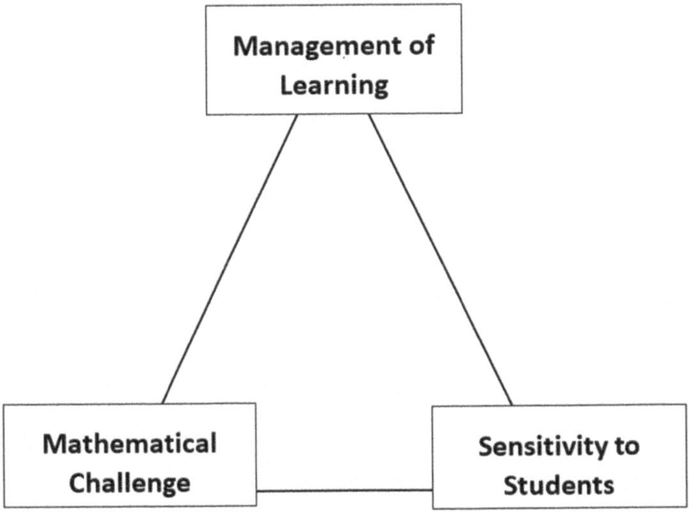

Fig. 5.3 A teaching triad after Jaworski (1996, p. 107)

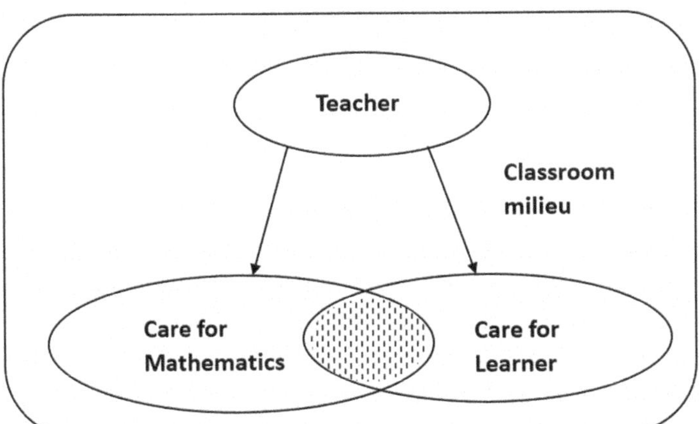

Fig. 5.4 A didactic triangle showing an overlap between care for learners and care for mathematics within the classroom milieu

taking care to construct what is to be learnt in ways that enable students to perceive it, connecting the intended 'object of learning' to the way it is enacted in the classroom, so that students' lived object of learning is as close as possible to the teaching intentions (Marton 2014). In the mathematical applications of Ference Marton's Variation Theory this connection is achieved by carefully coordinating perceived, or generated, variation so that students can attend to relations between variables. Comenius and followers of Maria Montessori and also Lev Vygotsky would insert the use of materials, material tools, gesture, diagram and language to further mediate the subject for the student.

These ideas are coordinated elegantly in Heidi Måsøval's thesis (2011) in which she points out that theories of learning in school have to support shifts from the spontaneous conceptualisations that arise throughout life to the scientific conceptualisations that are relevant for school-learnt disciplines. Typical shifts are made between: ordinal and cardinal meanings of number; multiplication as repeated addition and as scaling; angle as a measure of sharpness and as a variable in trigonometric functions; reasoning based on visual appearance and deductive reasoning based on properties. To make these shifts mathematics as a discipline has to be mediated by the deft use of tasks, tools, symbolisations, talk and students' own thinking. Guy Brousseau (1997) describes how teaching can support such shifts by devolving thinking to the students, and then validating and formulating the activity that follows through discussion and using conventions.

This kind of teaching requires sound mathematical knowledge; similarly, this kind of care for students requires sound knowledge of students. As Tim Cain (2018) points out, this might include intimate knowledge of local culture. In his experience, the research that a group of teachers found most useful was not about school subject pedagogy but about the lives of children in coastal towns—the children they were trying to teach. In many English schools, teachers are not embedded in the communities in which their students live. Jasmina, a secondary mathematics teacher, learns about her students' lives by volunteering to work with young people at a mosque. As well as learning about their friends, food and dependence on social media she also learns what life is like in nearby schools from their perspective. She also hears about the pressures, dangers and pleasures of their daily lives, including their fears about knife crime, self-harm, police action, drug dealing and how their own behaviour is constrained or disturbed by these. It is not surprising that against that background culture the majority of the young people she teaches are keen to get the examination grades that will give them employment and other opportunities. Some, however, seem unable to apply the efforts that would help them achieve. Instead of having traditional parents' evenings her school has a policy of holding individual meetings attended by herself, a more senior teacher, parents or carers and the student to discuss achievements, difficulties and ways forward. It matters whether the home environment supports education and the school and some parents and carers find it hard to carry out their part in decisions made at such meetings. Models of the connection between effort and achievement are, however, available in the school such as one student who had seven attempts to pass GCSE so she could become a nurse. One could question the relevance of GCSE content, the reasons for her initial failure, and the teaching between her first and seventh attempt, but what seems to triumph here is the focal flexibility of an adolescent as described by Coleman, deciding what to focus on to cope with competing tendencies and demands.

So far in this book several of Ernest's dimensions of mathematics education (1991) have been visited: politics as imposing curriculum and expectations, partly in response to international comparisons; the nature of mathematics and its complexity as a body of knowledge and also as a method of inquiry and thought; society with its disadvantaged groupings; childhood and adolescence and the pathways of development; learning as becoming more competent in mathematical thinking as well as memory; teaching being shaped by these dimensions. Learning mathematics has been seen as complex, involving different kinds of brain function, some of which have been mapped in cognitive science and many which have not.

I have presented examples of mathematical tasks that support the development of mathematical thinking by exploiting and developing inquiry habits that are not merely memory storage functions. There has been recognition of the roles emotional and social care play in supporting learning. I have offered three foci for care: care for mathematics in its fullest sense; care for students including knowledge of community; and these two are connected by care for their learning of mathematics.

In the three chapters that follow, relationships between care for mathematics, care for students, and care for their mathematical learning will be explored in a variety of educational spaces. In each space, the communities within which the spaces are located play a role. The teachers exhibit care in a variety of ways, not necessarily because it is imposed on them as a requirement of their work, but as an expression of the connection between their professional and personal being. In the next chapter, I report on the work of teachers in some communities that are extremely different from mainstream national cultures.

References

Ashcraft, M. (2002). Math anxiety: Personal, educational, and cognitive consequences. *Current Directions in Psychological Science, 11*(5), 181–185.

Boaler, J., Wiliam, D., & Brown, M. (2000). Students' experiences of ability grouping – Disaffection, polarisation and the construction of failure. *British Educational Research Journal, 26*(5), 631–648.

Brighouse, T. (2019). *Thatchers' market era is over in education.* https://www.theguardian.com/education/2019/nov/26/thatcher-market-era-over-education-election-hope-rival-attlee. Accessed 11 Apr 2020.

Brousseau, G. (1997). *Theory of didactical situations in mathematics* (N. Balacheff, M. Cooper, R. Sutherland, V. Warfield, Eds. & Trans.). Dordrecht: Netherlands: Kluwer.

Cain, T. (2018). *Becoming a research-informed school: Why? what? how?* London: Routledge.

Comenius. (1657). *Didactica Magna.*

Damasio, A. (1999). *The feeling of what happens: Body and emotion in the making of consciousness.* Boston: Houghton Mifflin Harcourt.

Davis, B. (1997). Listening for differences: An evolving conception of mathematics teaching. *Journal for Research in Mathematics Education, 28*(3), 355–377.

Ernest, P. (1991). *The philosophy of mathematics education.* London: Falmer Press.

Hackenberg, A. (2005). A model of mathematical learning and caring relations. *For the Learning of Mathematics, 25*(1), 45–51.

Heidegger, M. (1927). *Being and time* (J. Macquarrie & E. Robinson, Trans., English Edition). New York: Harper.

Hewitt, D. (1999). Arbitrary and necessary part 1: A way of viewing the mathematics curriculum. *For the Learning of Mathematics, 19*(3), 2–9.

Jaworski, B. (1996). *Investigating mathematics teaching: A constructivist enquiry.* Abingdon: Taylor & Francis.

Johnston-Wilder, S., & Lee, C. (2010). Developing mathematical resilience. In *Proceedings of the Annual Conference of the British Educational Research Association, 1–4 September 2010*. University of Warwick.

Marton, F. (2014). *Necessary conditions of learning*. London: Routledge.

Måsøval, H. (2011). *Factors constraining students' establishment of algebraic generality in shape patterns: a case study of didactical situations in mathematics at a university college*. Doctoral dissertation at the University College of Agder, Norway. http://www.researchgate.net/publication/334361695. Accessed 11 Apr 2020.

Merttens, R. (2001). The baby with the bath water. *Mathematics Teaching, 176*, 12–13.

Noddings, N. (1984). *Caring: A feminine approach to ethics and moral education.* Berkeley: University of California Press.

Noddings, N. (1992). *The challenge to care in schools: An alternative approach to education*. New York: Teachers College Press.

Noddings, N. (2012). The caring relation in teaching. *Oxford Review of Education, 38*(6), 771–781.

Potari, D., & Jaworski, B. (2002). Tackling complexity in mathematics teaching development: Using the teaching triad as a tool for reflection and analysis. *Journal of Mathematics Teacher Education, 5*(4), 351–380.

Roebers, C. (2017). Executive function and metacognition: Towards a unifying framework of cognitive self-regulation. *Developmental Review, 45*, 31–51.

Schoenfeld, A. (2012). Problematizing the didactic triangle. *ZDM, 44*(5), 587–599.

Walford, G. (2014). Academies, free schools and social justice. *Research Papers in Education, 29*(3), 263–267.

Watson, A. (2011). Adolescent learning and secondary mathematics. In M. Pitici (Ed.), *The best writing on mathematics 2010* (pp. 163–174). Princeton: Princeton University Press.

Wiliam, D., & Bartholomew, H. (2004). It's not which school but which set you're in that matters: The influence of ability grouping practices on student progress in mathematics. *British Educational Research Journal, 30*(2), 279–295.

6

Mathematics Education in Indigenous Communities

Why Is Indigenous Education of Interest?

In thinking about alternative mathematics education it is appropriate, as I have said before, to keep asking: 'alternative to what?' It happens that formal schooling orientated around the written word is how jurisdictions have developed ways to educate, but every individual child from whatever background has to learn for themselves how to 'come to know' in that kind of schooling, and it might be harder for students from backgrounds where such learning is not embedded. Rather than seeing this as a quality of different groups, mainstream cultures could see it as a high profile reminder about the learning of *all* students, whatever the background 'ways of knowing' in which they live, and that is why I have written this chapter.

I am going to look at mathematics education that takes place in alternative communities having traditional practices and forms of knowledge that are not from European heritage, and which seek to uphold these

---

The word 'Indigenous' is given upper case in the same way as 'European' would be given upper case. This is in line with current usage according to, for example, Canadian Press style.

A. Watson, *Care in Mathematics Education*, Palgrave Studies in Alternative Education,
https://doi.org/10.1007/978-3-030-64114-6_6

alongside the 'Western' values within which they are situated.[1] Mathematics education as a field can learn from these endeavours because the teachers in Indigenous contexts have to focus on attainment that matches national norms while simultaneously having to recognise that students have to navigate between overlapping cultures: community, nation and school, all with their own norms and histories. I have found also that the teachers in these educational spaces who are reported here draw attention to subtle features of mathematics that might be glossed over in a mainstream context. They often do this by listening both to learners and community Elders.

Qualities of care that I described in the previous chapter are written large as principles in the contexts I am going to describe: mutual respect; mutual listening; working alongside others with a shared focus; understanding what students already know; caring for human-made objects; caring for the inner circle of people and the cultural context. The work reported in this chapter is surrounded by supportive cultures and, in some cases, communities. Story is important as a form of knowledge. Language and talk is important, more so because of innate, often subtle, differences in meaning. The development of identity among school-age students, and their need for loyalty to some group, are recognised overtly in the teaching. In all contexts, the educational aims include, explicitly, the development of individuals who can navigate and coordinate various identities and this can, in part, be achieved through the teaching of mathematics. Teachers are carers, even healers, for this process.

In these Indigenous contexts heart, hand, mind and spirit/soul are often understood as components of everything that is human, including mathematics and the learning of mathematics. This kind of education sets post-colonial definitions and ownership of powerful knowledge against strongly collaborative communities who may have different views of knowledge and knowing. Mathematical knowledge can be understood differently, often being strongly associated with history, language,

[1] The term 'Western' is in common usage to express what are now global norms, generally promulgated through Anglophone agencies. For this book, I prefer the term European Heritage School Mathematics (EHSM) to describe mathematics as a discipline deriving from historical European work based on Arabic, Greek and Asian roots, with school mathematics being influenced by both the discipline and international testing.

environment and human activity. In Chap. 4, within a view of mathematics that I shall call European Heritage School Mathematics (EHSM), I have indicated a split between what has to be told or shown and what can be worked on through reasoning, perhaps together. European heritage has established common worldwide norms for school mathematics that are echoed in nearly all national curricula and all international tests. Briefly, EHSM starts with counting and learning four arithmetical operations, usually in this order: addition, subtraction, multiplication and division. Written calculations are privileged and tested. Alongside this, shapes are classified and named. The number system is extended to include fractions and decimals, followed by the introduction to algebra and graphs as ways of expressing general relationships. Reasoning with geometric and algebraic structures and generalities follows. Learning, performing and applying abstract, written, symbolic and precise methods, usually divorced from meaning outside mathematics, are the aim.

I write this chapter with humility. I am not a member of an Indigenous community nor have I ever taught in one, so all my knowledge is at best secondhand, often further away, and mainly gathered from publications. In general, teachers in Indigenous communities (like most teachers) do not write about the details of their work. Publications are usually in the form of glimpses of pedagogy, teacher guides and task sheets, or general reports about context, relationships with community and parents/carers. Occasionally, these have been written with Indigenous teachers who qualified within the prevailing culture and they have therefore learnt EHSM. Many researchers talk about background theories of equity and culture, but far fewer write about mathematical concepts. In the contexts I am using, governments and local communities are concerned about the consequences of underachievement, such as inability to participate fully in society as citizens and as contributors to the economy.

I began by reading published refereed papers about the work of some mathematics educators in Indigenous communities. I was aware that what I was reading were globally available research studies, often about special projects rather than normal pedagogic practice, and often written by non-Indigenous people. As my reading circled around through various projects I felt as if I was on a personal learning spiral, at the centre of which would be Florence Glanfield, Professor at the University of Alberta, who is herself Métis and a leading mathematics education academic whom I have known for several years. She would, I knew, have an overview informed by the mix in her own cultural background as well as her academic and professional experience.

Until my interview with Florence I had taken a Gramscian view that all young people need the high-order knowledge to which the groups in power have ready access through their own education or their purchasing power (see Chap. 1). This line of argument had led me to be suspicious of mathematics teaching that focused on everyday concerns in everyday culture since I could not see how these would provide access to more powerful knowledge. I believed that everyday foci tended to overvalue ad hoc methods for solving quantitative and spatial problems rather than providing everyone with the abstract tools on which further formal learning is based, and that shape quantitative power in the modern world. However, Florence reminded me that some extreme attempts to educate Indigenous children had, in the memorable past, removed them from their ancestral communities and denounced traditional ways of knowing. This was done in Canada and Australia and is still done in other parts of the world. Community Elders and keepers of traditional knowledge could justifiably see state-sponsored education as destructive, leading to dislocation and disorientation, with mental health issues arising from a damaged sense of identity and disrupted communities. Education, including mathematics education, should, she said, play a part in repairing this damage by respecting traditional knowledge and traditional ways of knowing. One way of going about this could be to shape mathematics education to be relevant for learners immersed in Indigenous communities while also giving access to formal mathematics culture, and a first step for me would be not to dismiss unfamiliar methods as ad hoc but as indications of different knowledge structures. I had read about 'culturally responsive' teaching but that phrase seemed to me to position educators as outsiders with respect to a culture, which many of them are, and who may also feel themselves to be culture-free. They are then urged to take the culture of the other into account when teaching EHSM. But none of us are culture-free. On the contrary, I would say that as a teacher I am embedded in cultures of conventional mathematics, English language and EHSM. If I am to be culturally responsive I need to be aware of my own culture as well as that of the student. Here is an example of how I am not culture-free: I know how to use the words 'sometimes' and 'often' in English because this is my first language and I have been immersed in its use all my life. However, an Iranian friend studying English as an additional language was told the approximate percentages that should be intended by words such as 'seldom', 'sometimes', 'often'—words I use with fluency. I found these percentages alien but not wrong. Being presented with systematic rules clashes with my cultural knowledge of how to use these words, but I can understand how guidance is needed for someone who has not developed that knowledge. The point being made by the phrase 'culturally responsive' is that teachers need to be aware that everyone brings cultural assumptions into the classroom. They do not leave them at the door.

A positive application of cultural response can be found in the work of Elle and Ovlla, two Sámi mathematics teachers who reconstructed teaching and assessment using local experience. They created contextual examination questions that contrast starkly with contexts those used originally such as airports, speedboats and parliament, but the mathematical abstractions in their work are the same (Fyhn et al. 2016). For example, reindeer husbandry needed measurement, number and algebra; Sámi clothes needed geometry and combinatorics; sledge construction needed geometry and measurement.

There is something revealing about cultural assumptions in a story from Edward Doolittle, a Mohawk scholar who is a university mathematician in the European tradition. Some educators were being shown how the four sectors of medicine wheel could be used to construct science teaching episodes. Medicine wheel is an image throughout North American First Nation teachings; it exists in the land, in images and in minds. Four equal sectors of a circle are sometimes said to represent spirit, body, mind and emotion; or unity, peace, harmony and courage; or four directions. It is understood to rotate, having no top, bottom or sides, and to be perfectly balanced. The leader of the session asked if the presentation of this structure for science teaching made sense. Elder Betty McKenna of Moose Jaw was asked what she thought about it. She replied simply: 'I have worked on a real medicine wheel' (Doolittle 2006, p.1). The imposition of a simplified structure onto her complex, embodied sense of knowing and being, did not impress her. A real medicine wheel is a dynamic map of the complex living relationships involved in being alongside other living and non-living things and people in the world, not an organisational tool for science teaching. Presenting it as a pedagogical organiser for school teaching was a well-meant misappropriation. In mathematics, Doolittle talks of oversimplification when well-meaning people try to bring aspects of someone else's culture into the purpose of serving abstract mathematics, for example, claiming that a tipi is a cone. This would be as inaccurate, gratuitous and insulting as using a typical English terrace house as an example of a cuboid topped by a prism, and also unhelpful because of the multiple social, psychological and cognitive associations a child might have with a tipi or a terrace house. These extremely different cases remind me that contexts can be misunderstood as well as misappropriated for mathematics and can have emotional associations.

Giving non-Indigenous teachers instructions about being culturally responsive has pitfalls. The Common Curriculum Framework for K-9 Mathematics: Western and Northern Canadian Protocol (Alberta Education 2006) describes Aboriginal students in its Program of Studies as having: 'a whole-worldview of the environment in which they live and learn best in a holistic way. This means that students look for connections … learn best when mathematics is contextualized and not taught as discrete components.' This description assumes homogeneity of Indigenous students, although they may have many different home languages and customs; some may have lived for a while in cities immersed in non-Indigenous culture, and many may also be exposed to digital media and the forms of self-controlled and selective learning that go with that. It goes on to advise that traditional learning takes place through activities with little emphasis in the written word. It is therefore 'vital' that teachers use authentic activity to communicate and enhance teaching and learning. This appears to me to describe out-of-school learning for all, not only Indigenous students. The people learning to skateboard or use mobile phones described in Chap. 3 learn similarly. 'Differenting' these kinds of learning from those of other Canadians might be seen as patronising, but the document was written to help. No one should claim that formal school teaching detached from the environment and daily life is a natural way to learn for *any* individuals, Indigenous or European—learning in school has to be learnt.

Preservation of traditional ways of knowing and coming to know is important. Additionally, there are conceptual details in many Indigenous languages that draw attention to distinctions that might be different from those made in Indo-European heritage languages, and different for different Indigenous groups. Having said that, it is also true that in my reading I have found remarkable similarities between mathematics educational issues in Indigenous communities in North America, Australia, New Zealand, Papua New Guinea and Finland. These will emerge as I describe some of the reported work.

Relevance of Mathematics and Context

It would be a mistake to generalise too soon that students who are marginalised from mainstream society, such as some Indigenous heritage learners might be, need some special kind of life-relevant mathematics in place of the high-status mathematics that is exemplified in EHSM. Renuka Vithal (2003), talking of South African black urban street young people, points out that their everyday lives often involve poverty, abuse, begging, slave labour as sex workers or drug runners, and they do not want mathematics teaching that brings their everyday worlds into the classroom. They want an education that leads to the qualifications that everyone values. But proponents of curricula that engage with their lived cultures are usually talking about schooling that takes place within communities that have shared values and ancestral knowledge heritage, not urban street contexts. The keepers of ancestral knowledge, and the carers of those young people, usually want education to include their own culture, and the language that supports it, alongside those qualifications that have external value and hence empower communities. Those of whom she speaks do not have that context. However, what students in Indigenous communities and the urban poor share is underachievement in mathematics.

Over the last three decades (at least), there has been much debate about the nature and importance of 'relevance' in teaching mathematics to underachieving groups of students. These concerns can be roughly categorised as:

1. Assuming that an underachieving student can only respond to knowledge that somehow arrives through an everyday context.
2. Teaching mathematics that is judged to be relevant to the students' present life concerns and experiences. This can trap students into poverty or a limited view of possibilities, thus reproducing divisions in society. Relating school mathematics to daily phenomena can over-mathematise everyday events, introduce unnecessary and inauthentic formalisations so that students who could function perfectly well mathematically everyday can be disempowered in school. At its worst, cultural artefacts are used gratuitously, a position that Narcisse Blood satirises as: 'Throw a bead and feather in there to shut them up.' (Sterenberg et al. 2010 p.8).
3. Entwining culturally specific mathematics with EHSM in the curriculum so that students can appreciate the connections between. The work of Elle and Ollva indicated above illustrates this. Sometimes mathematical ideas can be found embedded in the culture and be used as a legitimate basis for the formalisations of standard mathematics, such as abstracting patterns used in weaving, braiding and tiling as mathematical structures, and being honest about the abstraction process. Alan Bishop (1988) points out that all cultures have ways of counting, measuring, locating, designing, explaining, and playing that could be, or could become, or could be expressed as, mathematical. Generalising and formalising these by developing symbolic representations (e.g. developing measuring tools; using musical notation) is a human act allowing communication between people and transfer of knowledge between contexts.
4. Excluding culturally specific mathematics that does not match EHSM and presenting EHSM as a separate, school-based endeavour.

In most of the Indigenous contexts I read about it was clear that Elders wanted young people to learn EHSM mathematics alongside learning to be a knowledgeable member of their traditional community. By contrast, Sámi Elders are less likely to be supportive of EHSM aspirations since the major concern is sustaining their way of life so avoiding engaging with anything that is not directly enabling life in a challenging environment (Fyhn et al. 2011).

Entwinement

In all meanings of 'relevance' above it matters who decides what is relevant. Some of these stances slither towards teaching only the mathematics that the authorities imagine is going to be relevant for certain social groups in terms of their imagined futures. A tendency to teach students from historically underachieving groups only the procedural arithmetic as useful for everyday life is noted in the PISA international test findings (DfE 2016) and I see this also in some classes in England. A frequent underlying assumption is that Indigenous mathematical worldviews are simpler than EHSM—which is often not the case as my earlier attempt to describe medicine wheel indicates. In Australia, Helen Verran compares the Western system of land measurement by acreage and position to a traditional Yolgnu Aboriginal system in which the measurement is a vector, understood in relation (Verran 2000, p.306). Such a matrix includes relations of varying intensity, direction and clan relations; knowledge about it, the intellectual work, is learnt through story and song. Hardly a simplified method of mapping land ownership!

Guida De Abreu and David Carraher (1989) describe a traditional method for calculating field areas among sugar cane farmers in Brazil. Individual farmers need to know how to make these calculations, but the methods are culturally embedded rather than taught in school. First the field, that is usually an irregular shape, has to be approximated to a composition of quadrilaterals and triangles. Then a calculation based on lengths has to be made, the length having been estimated using a measuring stick that is approximately a standard length. The mathematical principle of measuring by repeated abutting of units is understood through its enactment. The lengths of each pair of opposite sides of the approximate quadrilateral are averaged and the averages are multiplied. A triangle would be treated as a quadrilateral with one side of zero length. The process of averaging is described as the longer side 'lending' length to the shorter side. At every stage numbers are rounded to integers. The area of a quadrilateral ABCD is $\frac{(AB+DC)}{2}\times\frac{(BC+AD)}{2}$ approximately.

This method of describing areas of fields can be directly compared to the more precise methods taught in EHSM settings to trigger a discussion about the scope of its application. As with many local procedures this comparison will reveal that, with the range of fields to which they are likely to be applied, more precise methods might be unnecessarily unwieldy in the context. Recognising the strengths and limitations of the local or craft methods provides students with the power to discern their usefulness. Students would need to be convinced of a need for alternative methods if they were immersed in a culture that functioned very well with local approximate methods.

Relevance, therefore, has a complex part to play in mathematics education. I am using the idea of 'entwinement' to avoid the divisive nature of 'biculturalism', and also to avoid hierarchical assumptions from the perspective of students. Thelma Perso (2006), working in rural communities in Queensland, likens this to the flow of two bodies of water into each other. Tony Trinick, Tamsin Meaney and Uenuku Fairhall (2017) have suggested 'cultural symmetry' as an approach to entwinement, namely to discuss the design and use of artefacts from the two perspectives of mathematics and tradition.

Community Support for Mathematics Learning

The situations that Florence and I discussed are those in which learners are brought up in established Indigenous cultures that differ from European culture, particularly where the community wishes to uphold traditional views, while recognising the power of other worldviews. Issues in Māori, Sámi, Australian Aboriginal and North American First Nation education are similar although the histories of imposition of European education are slightly different.[2] This involves alignment of language and

[2] Although Māori were not prehistorically in New Zealand they are regarded as Indigenous in that they were there before European domination and have a strong culture that is maintained by community leaders. They face similar problems with EHSM education as other Indigenous groups because their worldview is not European. They have traditional ways of seeing quantities and spatial concepts and have historically been judged as less intelligent because of these differences in foundational concepts.

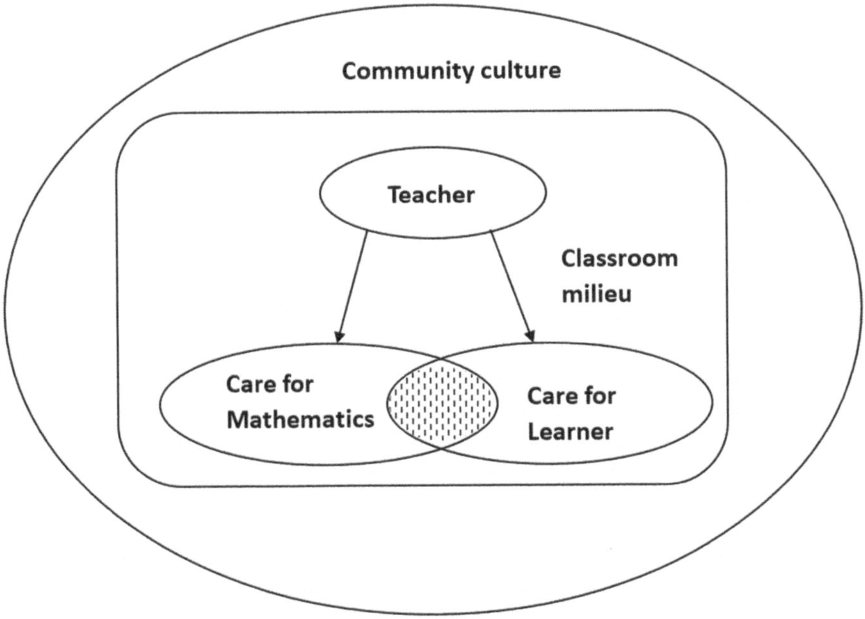

Fig. 6.1 A caring didactic triangle within a supportive community culture

also classrooms and schools becoming culturally authentic spaces in terms of their organisation and social interactions. Care by the community for students and for mathematics is strongly present and suggests an embracing feature that might be added to a didactic triangle (see Fig. 6.1).

Community and Māori Mathematics Education

Uenuku Fairhall is a Māori mathematics educator who has been involved for many years with ways to entwine traditional and EHSM mathematics education in one school, Te Koutu. This work has contributed to the development of a national mathematics curriculum that respects both original worldviews, Māori and EHSM, an example of entwinement. The first version of this curriculum focused mainly on translating EHSM mathematical language into Māori. Bill Barton, Uenuku Fairhall and Tony Trinick (1998) undertook 15 years of work, along with the Māori

Language Commission, constructing culturally appropriate language for many mathematical ideas. It seems they were relatively successful: for example, a word they constructed for calculus from fundamental concepts in Māori language turned out to exist already with the meaning 'to move with an even motion', thus matching ideas that arise in the early use of calculus in well-behaved contexts. Learning mathematics in two languages that have two sets of cultural foundations can be helpful because it can give two different, but mathematically equivalent, perspectives on an idea or phenomenon. The language work was done with full participation of Māori Elders and parents who wanted the etymology to be Māori while also conveying EHSM conceptual development. The same research team found, when working later in the same school, that community members and teachers all recognised that mathematics was embedded in many of their common practices:

> We go to the marae [meeting ground], there's their mothers and fathers in the back there, doing all the menial tasks of getting kai [food] ready. There's measuring, cutting up, those are all maths-type things, in their everyday life (Meaney et al. 2013, p.246).

A further example is that bilateral spatial symmetry is understood from the structure of fishbones, the shape of the iconic fern leaf, the human body and also in traditional decorations and carvings, so has meaning beyond both everyday and mathematics as something that is fundamental in the natural world. It is not, therefore, a gratuitous context to be experienced through a textbook but something to be lived through actions.

Community and Yup'ik Mathematics Education

A team led by Jerry Lipka, a non-Indigenous educator, illustrates what can be achieved when an outsider puts human relationships and respect in the forefront of work to raise educational aspirations and achievement. His main work is the *Math in a Cultural Context* project (MiCC) with Yup'ik people in Alaska, extending over three decades (e.g. Lipka et al. 2019). The mathematical work I summarise here arises from listening to traditional knowledge holders and following their conceptual insights. In common with many Indigenous people bilateral symmetry is a key idea, together with senses of balance, equality and measuring-as-comparison that arise from symmetry. To this is added, in Yup'ik culture, the notion of centre, being the crossing of two orthogonal axes of symmetry. Lipka learnt that centre, symmetry and measure are embedded in many daily activities, often in combination with each other such as when halving by folding. Building on these ideas, working all the time with Yup'ik Elders and educators, his team developed a mathematics curriculum that grew from these fundamental perceptions, rather than the counting and naming that underlies EHSM (Lipka et al. 2019). Yup'ik counting is a two-dimensional activity based on four directions, so that the equivalent of 'number' has origin, orientation and measuring unit. It is a small step of the imagination to see that this system can relate to scaled orthogonal cartesian axes or vectors. Representing numerosity with length, similar to using Cuisenaire rods, presents additive relationships implicitly, so that $a + b = c$, $c - b = a$, $b = c - a$, and so on all coexist. This could be an advantage in early algebra. It might not quantify in the ways EHSM does, but instead gives relative position and hence distance. It also gives obvious visual access to linear functions by observing $f(x) = x$ and $f(x) = -x$ and then transforming these through movement to generate other functions.

Yup'ik knowledge holders can fold strips of whatever is available into successive halves to construct measuring tools and they practise extending this process to very accurate comparisons of length. The activities of finding a lowest common multiple or highest common factor to compare lengths, when one length cannot exactly measure the other, naturally follow, and multiplicative relationships and ratio become available as ideas. Again, an imaginative leap relates successive folding to a binary system based on powers of 2 and also to exponentiation. What if powers of two cannot help? Other divisions can be made by successive halving of parts of the strip while shifting the axis of symmetry to different parts of the strip if necessary. In EHSM mathematics these would be described as finding fractions and the folding method would be called 'iteration'. To fold a strip into five equal parts, estimate one part at the rightmost end and then fold the remainder in half twice to make four equal parts. You now have four equal parts and one estimated part. The part on the

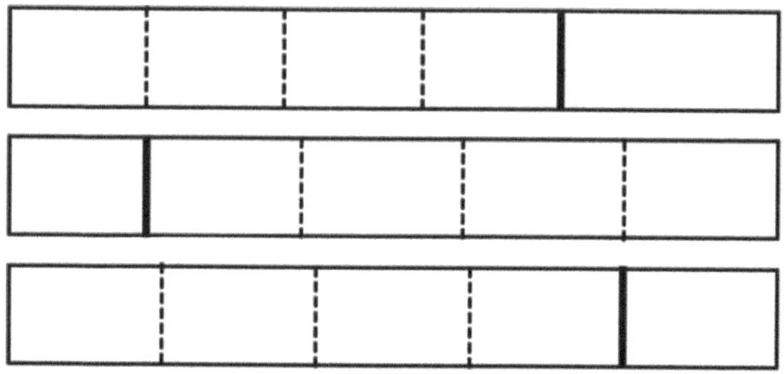

Fig. 6.2 Folding fifths (the dotted lines are the fold lines at each stage and the solid line indicates the segment omitted from the folding at each step)

leftmost end is the next approximation. Ignoring that, fold the rest of the strip into four equal parts. The part on the rightmost end is the third approximation. Continue in that fashion until all five parts are equal (Fig. 6.2 shows the first three folding states).

In practice, fractions can be achieved closely with very few iterations. Yup'ik measuring can be as accurate as it needs to be, so accurate that Lipka et al. report that:

> When Dora Andrew-Ihrke went through this process to compare the circumference to the diameter, she used her folding algorithm to determine what was left over after 3. Once she reached a point, she said, "it's an ugly number." She commented later: "first time I couldn't measure!" (Lipka, Adams et al. 2019, p.141)

Dora had worked with Lipka for a number of years, drawing on the knowledge of her mother, who was an Elder. In EHSM it is known, but hard to demonstrate, that π can never be a finite sum of fractions.

Changing the curriculum is not the only feature of teaching in the schools that have been involved in Lipka's work. Their pedagogy also aligns more often to traditional ways of learning through expert-apprentice methods with adults and children working alongside to make things together. The emphasis on symmetry is evidenced in the teaching of Nancy Sharp (Lipka, Sharp et al. 2019). She led a shared project to cut paper to make the traditional border patterns that decorate parkas. While this went on there was conversation about the shapes, their orientation, their characteristics, congruence, similarity, all embedded in the authentic activity. Making the patterns, with the associated talk and teaching style, is similar to a report of another teacher, Cap, who was teaching the traditional craft of making beaver pelts (Lipka 1991). Lipka noted that the beaver-pelt lesson did not have a lengthy instructional introduction; instead the necessary 'instructing' emerged from the teacher's comments as he made a pelt, and the children's own comments. The learning in these examples is through apprenticeship that is totally appropriate as the child is learning to be Yup'ik alongside an 'expert'. A representation of this as a didactic triangle would have to show the teacher and student alongside (as in Fig. 5.2) with the insertion of tools and cultural norms as the ways mathematical ideas are mediated for learning (see Fig. 6.3).

Although Lipka's methods showed measurable educational benefits in learning about numeration, measurement, data handling and place value when rolled out across many schools, the authors point out that it is not possible to disentangle the effects of professional development, cultural materials, the problem-centred pedagogy, and engagement with research, to say what each factor contributed to successful learning. For example, in the task 'Going to Egg Island: Adventures in Grouping and Place Values' a Yup'ik girl makes her first trip to gather eggs using tools and

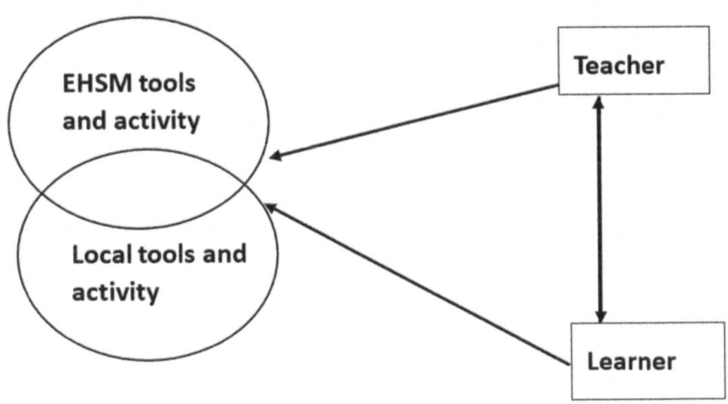

Fig. 6.3 An 'alongside' didactic triangle with interaction of local and EHSM mediation

everyday methods for her journey; in the task 'Building a Fish Rack' students design the rack and determine the drying area for various shapes and their structural stability. The contexts are authentic, the necessary collaborative work is authentic, and the mathematics used in them is also authentic but addressed much more explicitly than would be usual, with mathematical language and symbolism. Entwined stories about curriculum, pedagogy, community engagement, mutual learning and explicitness about mathematics can be seen in these examples (Kisker et al. 2012).

In practice, teachers in roll-out used a mixture of methods. One teacher, Stacey, herself an outsider, inserted EHSM ideas of proof (Lipka et al. 2005a, b). To establish these as a norm in which students could lead discussions she constructed an enclosed safe learning space for Yup'ik and EHSM methods to co-exist.

Community Learning in Northwestern Ontario

Joan Moss, Bev Caswell, Zachary Hawes and Jason Jones (2017) linked spatial reasoning and general mathematics progress in a programme that has reached over 2000 early years children across Ontario. This *Math for Young Children* (M4YC) programme came to the attention of Elders of Ojibwe communities in Northwestern Ontario and the team was invited to introduce it to Indigenous teachers there. When a spatial approach to mathematics was explained, Mike Kabatay, an Elder, commented that Anishinaabe people did not think of this as mathematics, since it was their 'way of life'. Partnership with all concerned, Elders, parents, carers, local teachers and education officials, was crucial to this success. Their goal was to form a collective in which all partners could contribute expertise equally, not to mathematise local practices but to entwine them with EHSM ideas. Collaboration was such an important feature that Jason Jones, the Ojibwe language coordinator for the district, created a word: *Gaamaamawiasigagindaasoyang*, meaning 'gathering to learn and do mathematics together'. The project was successful for children's learning across the whole of mathematics, sometimes beyond the usual level in EHSM curriculum.

Some tasks were created specifically for the community context, incorporating mathematics. For example, in a special event, families collaboratively created their clan animals from tangram shapes. Older students had made wooden fabric printing blocks of the same clan animals and families used these to print squares that were then sewn into quilts. Other tasks will be familiar from other resources, such as finding the 12 unique shapes that can be composed using five congruent squares and comprise the full set of pentominoes. This is a considerable challenge for very young children who have to apply a concept of congruence of composite shapes while also visualising transformations of rotation and translation. In the second part of this lesson, children had to visualise and then justify which of the 12 pentominoes, when folded, could make an open box—thus adding more visualising, in this case from two- to three-dimensions. All this work was accompanied by discussion and listening to the students as they explained their ideas.

As well as exploratory tasks that sometimes drew on cultural practices there were quick spatial puzzles (Moss et al. 2016). These provided a context for using prepositions and the language formats of EHSM mathematical reasoning. One was to copy a 2-D object and extend it by reflective symmetry, using precise language to explain how this was done. Another was to give instructions for replicating a shape that had only been glimpsed. Another was to describe the edges, faces and orientation of a 3-D objects. All these tasks involved symmetry, balance, equality, structure—key ideas in both mathematics and Indigenous worldviews—and through them the use of mathematical language was developed.

Number Differences

A significant difference between some cultures is the meaning of number. In EHSM counting is often used to denote accumulation, whereas in some Indigenous cultures counting is more often used to share by grouping of subitised amounts; this action has more in common with multiplicative than additive reasoning.

Tamsin Meaney and Deb Evans (who is herself Aboriginal) (2013) refer to Mary Gilmore, an Australian writer who learnt to count stars alongside Wiradjuri children. She learnt to recognise patterns of numbers as units and then count the units and groups of units and so on. Counting anything in groups of three by subitising became her natural way of counting—a number system that does not require any more number words than the base group size. Zoltan Dienes used this idea of groups of groups of groups in his design of number base blocks (1969). His 'base ten blocks' help children learn about place value in base ten.

In some number systems the basic pattern is five, the word for five being 'hand' or 'foot' (Hunting 1987); there are derivations of this idea in the Indo-European language group (the symbol and word for five being like a fist) and also in the construction and operation of some abaci. Lipka and his colleagues report that Yup'ik Elders made four bundles of five sticks that were then regrouped into a complete set (Kisker et al. 2012). Therefore, in their teaching they built on base 20, subbase 5, to establish base 10 place-value.

A project in the Kimberley (Australia) had some success with teaching the EHSM number system keeping a tight conceptual focus within games or collaborative tasks and appropriate teacher interactions (Jorgensen et al. 2011). They used the physical familiarity of 5s in a variety of number and money tasks. 'Counting back' was very hard and they used games that focused on this, for example, going from ten to zero. Where fundamental understandings of number are different, then the early stages of EHSM can be confusing and some children may be judged as having deficit, rather than different, knowledge, yet the connections I have just described do not come naturally to children learning in EHSM either—they have materials, chanting and practice to embed these ideas as additional to what they learn in rhymes, daily experience and games. A significant language mismatch may have occurred in the Kimberley project when the teacher tried to use a numberline. Imagining how many linked cubes were hidden in a tower made of 15 cubes offered a spatial image for subtraction; finding pairs of numbers that give the same answer when subtracted offered instantiations of 'difference'. In common with learning mathematics everywhere, spatial understanding was used as a foundation for mathematising but the local language of Nyikina made different distinctions from those in English: they had 34 distinct terms for direction and location and no words for 'right' or 'left'. It is possible that the numberline as an object did not link easily to their vocabulary, and so did not function helpfully as a representation. Shifting between representations in general, including spoken words, is a different experience in languages that have different foundational concepts and in cultures that have different spatial and symbolic ways of knowing. Connections assumed in EHSM between spatial and numerical concepts, via measurement, can be a source of confusion. In Sámi culture, there is a word that translates as 'half' but it can mean 1:2 or 2:1 according to the context being measured or quantified (Fyhn et al. 2011). Students therefore have to adapt their use of it according to material context, but also according to whether they are at home or in school. Differences between home and school language apply in English-speaking homes of course since 'half' does not always mean 'two equal parts'.

Ways of Knowing

Bill Barton only realised late in the Māori curriculum work that numbers are more like verbs rather than adjectives and nouns as in English. This is because 'numbering' or 'quantifying' are actions. It is striking that this verb-like quality of number is present also in reports from North American and Australian educators who come to understand how Indigenous mathematical language differs from EHSM. Enumerating using verbs rather than nouns could mean that in some languages it is difficult to make sense of ordinal numbering. Robert Hunting (1987) elaborates this observation in the case of Yolngu counting systems. You could touch the first object saying 'one' but you cannot touch the second individual object and say 'two' because that is only one object. You can use 'two' to describe them both together. In EHSM, the dual use of number words as ordinal and cardinal has to be learnt. I remember a three-year-old English child asking her father, a mathematician, 'What's three about this one?' when counting three chocolate biscuits by touching each. In EHSM, the tendency is to start with the words and one-to-one correspondence so it is not possible to know whether it is more natural to 'see' groups with a certain number of components through their spatial arrangements and hence subitise their quantity.

As well as the different meanings of number words described above the use of verbs where EHSM uses nouns goes further. Marjolaine LaPointe, a teacher from the Deer clan working in the M4YC project described below, said: '… most of our words are verbs so when we do math, we're not doing math, we're mathing!' Lisa Lunney Borden (2011) worked for many years teaching in a Mi'kmaw community, learning the language, becoming immersed in community life and developing mutual respect with the Elders and the parents of the students she taught. She writes about her realisation that verbs are important to describe what happens rather than labelling objects. Through working in the We'koqma'q First Nation she came to understand the need not only to learn Mi'kmaq language but to learn *from* it. In her writings, there are several examples of how she adapted the way she taught EHSM concepts to the students' commitment to actions rather than objects using visual, physical, and tactile approaches. One way would be to use familiar local contexts, for example introducing proportionality through comparing the measurements of canoe paddles of the ottertail design for adults and children: the ratio of measurements between and within paddle sizes is close to constant. A more subtle shift was to change her questioning to match their perspectives, so rather than asking 'what is slope?' she would ask 'how is the graph changing?', a shift from noun to verb that additionally (though she does not mention this) draws attention to the covariation of the variables that is the essential relationship being graphed. EHSM language created tensions for her Mi'kmaw students who 'sought actions but were presented with things' (Borden 2011, p.11).

Reading various reports of 'verbification' highlighted for me a difference between school mathematics as logical, framed by rules, facts, methods with a structure of noun-like concepts and a mathematics in that a sense of relating elements of quantity and space comes first and the rest follows. I have long held both views and had not seen a problem with that, but my interest in relationships had been abstract. For me, for example, the relationship between, say, distance travelled and a wheel becomes mathematical when I can describe that relationship in terms of varied distances, rotations, positions, rates of change (velocity and acceleration) and so on—the work of mathematical modelling. But I now see that a classification of concepts according to the behaviour that manifests them relates more closely to human experience.

Another feature of these traditional worldviews is that mathematical knowledge is evidenced around us, even the abstractions of their views are present in all things. Mathematics teachers enact this when they reach for models that students can see or imagine in order to demonstrate abstract principles. Abstract ideas such as balance, equality and structure might exist in traditional cultures, and each of these ideas would be typically supported with some kind of story that holds within it some essential relationships. But rather than Indigenous people being pushed into EHSM straitjackets, or well-meaning teachers shrinking EHSM to fit what they imagine to be Indigenous everyday thinking, Doolittle and Glanfield (2007) would like each to feel the pull of the other, so that balance can be achieved. While writing this book I came across the idea that inequality and imbalance are important too, because some mathematical ideas are about restoration and maintenance of balance: negative numbers; manipulating equations; symmetry; moments; complements.

The Kimberley Project understood the importance of two-way learning between EHSM culture and local cultures, for example that learning traditionally takes place through language, story, song and 'yarning', that is a dialogue process in a circle during which people can share their experiences, thoughts, questions, stories and meanings. Their pedagogic aims were less successful; the import of a particular groupwork pedagogy clashed with certain traditional collaborative practices and many teachers could not use it in ways that were also mathematically rigorous (Jorgensen et al. 2011).

Traditional knowledge keepers do not regard formal mathematical knowledge as of a higher order by being rational—rather they see that EHSM has lost part of the holistic unity of mind, heart, spirit and body. Ironically, the experiences many EHSM mathematics educators hope students will get from doing mathematics include awe, wonder, excitement, puzzlement, satisfaction and a sense of personal power, that are emotional and transcendental sensations, yet in EHSM these facets have been separated in order to measure knowledge and hence be accountable to political powers—awe and wonder are extras rather than the source.

One exception (there are probably many others) to the perceived need to gain EHSM qualifications is the *Sistema de Aprendizaje Tutorial* (SAT) approach to secondary education that has spread among marginalised rural communities in Latin America and some other continents. Rather than gaining internationally recognised test results the aim is to become more informed and powerful within the home community. The example I know of is used in Honduras (Robinson and Winthrop 2016) but the same principles are used in other countries, most successfully in Colombia where it originated. The communities in Honduras are mainly Mestizo (mixed race) with smaller numbers of Mayans and other Indigenous groups. Community Elders understand the need for mathematics and other forms of education but, like Sámi leaders, want to be connected to their economic and cultural survival and empower them to continue to live as they have for generations, which in their case is by farming. Influenced by the educational thinking of Paulo Freire (e.g. 1972) the schooled knowledge is related to the way of life. Secondary education for girls is highly regarded and 80% of SAT students in Honduras are girls, the boys being mainly already out at work. Pedagogy is project-based with textbooks that map integrate different disciplines. Teaching is undertaken by trained tutors, each having roughly 30 students, who are expected to learn alongside their students. The tutors, many of whom are from the communities, live in the same conditions as their students, all involved in growing crops alongside parents and other family members. Learning is through activities in the community and formalising knowledge takes place in discussion, seated in circles. As well as tasks that directly relate to what is going on around them, such as understanding insect populations, health issues and so on, they engage in thought experiments about situations further afield. Moral, social and democratic features are included alongside business and economic matters. Indigenous knowledge, particularly of plants and their medicinal uses, is included in many communities and their festivals. The development of useful capabilities is more important than disciplinary knowledge but nevertheless about 17% of those who do the course go on to college and some to degree level studies. This is often from a starting point of limited and inconsistent, if any, elementary education that is likely to have mainly involved rote learning. In mathematics (Richards 2010), there is some use of traditional systems and methods. Sarah Richards (private communication 2020) comments that tutors who knew more EHSM were more likely to adopt direct instruction rather than the expected methods of immersion in integrated projects; however, she adds that:

> they were much less directive than they would have been in a traditional school. It seems to me that people who are good at maths are more likely to judge their own teachers positively and to try and emulate them. This bias makes them more resistant to learning different practices.

Notes from Lochville

Now I look at an urban school in a large Australian city where Indigenous students are disconnected from traditional roots and contexts and therefore lack the embrace of a community that carries their cultural heritage. School, therefore, has to develop 'belonging' and attunement to EHSM learning with disenfranchised students. This provides a bridge to thinking about alternatives in England in the next chapter.

Lochville elementary school serves an area in which many young people identify as Aboriginal from a wide variety of language groups. Some of them have an Aboriginal home language and cultural context and some do not and know nothing of their heritage. Many live with carers who are seldom from the same cultural group as the child. The Aboriginal Education Consultative Group (AECG) works with the community, giving advice to schools about how to meld Aboriginal culture into the school where European culture is the dominant shaping force in the education system. 'Melding' is not enough, because there also has to be a restorative, healing role for education. Several generations had their traditional knowledge and ways of knowing deliberately removed by earlier education policies. Although parents and carers value learning and recognise the need for school-learning in the modern world, they can also see schools and the system as an alien and potentially damaging environment that removed their own cultural practices. School teaching took away their culture. Lochville therefore sees one of its roles as repairing this damage and restoring cultural identities, with the help of AECG, by reconnecting learners with their people and history. The whole school takes part in this. The school has an Aboriginal headteacher and other Aboriginal staff who are passionate about supporting healing and change.

Children are educated about traditional culture from the beginning of school, immersing their learning in everyday environments that transcend the immediate urban context. Many learning experiences are outside the school buildings. They also learn how to learn, both in traditional ways and in school ways. Teachers do not expect most learning to be done by children listening; instead they expect children to learn by watching, exploring, using physical resources, seeing others model actions and behaviour. They learn in collaboration with each other and, like any children, they enjoy working together. Teachers also identify areas of interest to students and use these as contexts for discussion of mathematical ideas.

A significant difficulty is the role of 'shame' that also affected the Kimberley project described earlier. This pervades traditional Aboriginal culture and has lasted beyond loss of other traditional ideas. I do not think it would be right to assume this is exactly the same as for European students for whom experience of being criticised or ridiculed by others has created fear of being seen to be either wrong or nerdy. Instead it has been a cultural norm for generations to avoid both the shame of being wrong, and also to avoid shaming others. It can be associated with respect for teachers as knowledge holders and avoiding disappointing them. It also ensures that the group learns together with no one being singled out by status. This can lead to a student shutting down completely and not even trying to do a task or give an answer. In mathematics lessons, there are probably more opportunities to be wrong than in any other subject. Avoiding asking closed questions can help, but doing mathematics always involves over-generalisations, miscues, partial conceptualisations that might turn out not to be appropriate, trial-error-adjustment routines, and following alternative paths of reasoning. Students therefore need to learn that it is part of mathematical learning to make unfruitful moves or misapply reasoning and that it is acceptable to, sometimes, do so.

The school navigates and coordinates the curriculum, identity, emotional and cultural foci of their work. Parents and carers and the AECG are as much involved in the planning of teaching and learning as the teachers and the students. This is achieved by yarning with adults and carers in a comfortable place to construct a personalised learning pathway for each child. In this space, among other things, the importance of number and numeracy will be discussed and shared goals for a student will be constructed. This is the space in which the interests and cognitive progress of a student inside and outside school will be braided with the demands of the curriculum so that everyone agrees and understands what is achievable and how it can be achieved over the next few weeks. This is an immensely demanding, but rewarding, task for teachers and sets the scene for future schooling and for the autonomy of the student in setting goals, managing and reflecting on their learning. It is also the space in which the understandings of parents and carers about the education system can be transformed.

So how does student individuality relate to the teaching of mathematics? Teachers have undergone a radical change to the way mathematics is taught in the school. It used to be a matter of producing and monitoring different levels of worksheet, differentiated by task, with support as and when needed after some teacher talk about how to do things and what they meant. I had an impression of lessons with a focus on procedures rather than meaning. Recently they have joined a long-lasting project that shifts them towards a problem-based approach in which the focus is on building concepts rather than memorising methods. They do this through collaborative work, often with materials, rather than individual paper-based work, using open-middle tasks rather than closed procedures. The educational aim is not to learn a sequence of things to do with numbers and shapes, but to develop a range of skills and techniques for solving problems with different levels of efficiency. The philosophy of the mathematics classroom is about teachers listening to children, taking turns, dialogue among children and between children and teachers, sharing multiple strategies, engaging in productive struggle and being given, and accepting, challenges. Students are empowered and supported to unpack problems and decide directions of effort. The mathematical work is at first divergent, but is closed down to mathematical convention through dialogue, identifying efficient methods 'that a mathematician would use', provision of conventional language and symbolism, and relating it to each student's pathway goals. They use a pedagogical framework: launch, explore, summarise, reflect, apply or analyse. Much of this mirrors traditional ways of learning and recognises individuals in a shared educational journey, which is why they welcomed engagement with the project.

Significant time and attention is given to early years' everyday experience of continuous quantities such as time, distance and growth as an experiential basis for becoming numerate. This lays the foundation for the kind of mathematics that is not about counting and calculations with integers, but is about ratio, estimation, multiplicative reasoning, spatial reasoning, relationships and patterns, while also being a context for reconnecting with traditional knowledge of weather, seasons, journeys and so on. It is important to note that awareness of these ideas is foundational for all children learning mathematics anywhere, and this project was not conceived specially for those with Aboriginal backgrounds. The whole number system is a tool for expressing these comparisons and relationships. In EHSM, everyday quantification is often under-appreciated as a basis for mathematical understanding. Ratio, for example, seems to be something people can understand in the context of their daily lives but in EHSM it often appears as a progression from multiplication and division seen as repetitive addition and subtraction and hence associated strongly with calculation rather than physical and visual comparisons.

Here is one example of teaching in Lochville for year 1: Students have past experience of subitising, and of estimating about 40 to 50 counters. They visit clear plastic token boxes that are placed in the playground containing tokens that are stacked up but each token is visible. As a whole class, they are told that they need to know how many tokens are in each box, as this is going to lead to attendance at an end of term party, by counting them without removing them from the boxes. They have mini-whiteboards to record their counting and they work in mixed groups. They are encouraged to say what is in their heads about how the tokens could be counted, and why some strategies might be more efficient. Anticipated responses include counting by ones, trying to count a small portion and then estimate. Students may well see something different to each other and to what the teacher sees. Discussion can then follow saying what they think about other groups' ideas about counting. The teacher listens to conversations in groups and finds a pair or group who are using a more efficient strategy, for example, grouping into 2s, 5s or 10s.

They are asked to explain what they are doing: 'Why is this better? How can you keep track of the count?' Someone may notice tokens could be in stacks, but if not, student attention is drawn what they can see visually: 'How are the counters arranged? Can you use this to your advantage?' They then make a class list of efficient ways to count a large collection.

Challenging questions are used to support students to move from simpler to more efficient approaches. For example:

How many so far?	Is counting in 2s more/less efficient than in 5s, or 10s?
How can you be sure?	Can you record how many tokens there are?
Where are you up to?	Would organising the counters in a certain way help?
Did you predict?	Can you prove it without counting individually?

Children experience that there are different ways to count and that objects can be arranged and sorted in different, and unexpected ways. They then explore ways to determine other strategies using mathematical tools such as weighing.

This lesson does not separate social, emotional and linguistic components of learning from cognitive aspects. Although not all students might achieve secure understanding of the value of counting in 2s, 5s or 10s and the relationships with finger counting or multiplication facts, there has been common experience that establishes episodic memory from which more learning can develop. Additionally, there has been some sharing, not as right and wrong ideas but as the raw material for a discussion about efficiency and what that might mean. The notion of 'being sure' is introduced even for these very young students. It also involves mathematics outside, groupwork, and an opportunity to go beyond 'shame' feelings. There could be an element of yarning in the final stages of such a lesson to think about that ideas can be carried into the next lesson.

What About the Teacher?

There are some imperatives for all teaching that arise from the commonalities in this chapter, but first I go back to Florence to talk about the importance of relationships. An important relationship so far omitted from my writing is the teacher's relationship with mathematical ideas. Florence starts with herself, remembering her own experiences and perceptions of a mathematical idea and moves outwards to her relationship with mathematics and with students, and hence with their mathematics:

- How do I understand multiplication of two binomials?
- In what way(s) is the multiplication of two binomials related to multiplication of two two-digit numbers?
- How can I invite my students to see the relationship between the multiplication of two binomials related to multiplication of two two-digit numbers?
- What were my experiences in learning about the multiplication of two binomials?
- How could learning the multiplication of two binomials contributes to my student's deeper understanding of mathematical ideas?
- In what way will my actions as a teacher in teaching about the multiplication of two binomials sustain my relationship with my students?
- In what way will my actions as a teacher in teaching about the multiplication of two binomials sustain the relationship my students are developing with mathematics?
- In what way does how I see myself as a student of mathematics impact on the decisions I make as a teacher of mathematics and the tasks that I ask my students to engage in when learning about the multiplication of two binomials?
- How do I participate in the development of a shared understanding of these mathematical ideas with the community in which I teach? (adapted from Sterenberg et al. 2010)

It is clear from these questions that the underlying 'answer' to this set of questions is not a mnemonic for multiplying brackets to be remembered for an upcoming test, but something much more complex about caring relationships with, through and around mathematics. It will probably be more like a conversation than a worked example and a list of examples to work out, however efficiently designed. She cares for their learning by caring about how she came to know and how they might come to know, through relationship with her. This is what I mean by cognitive care in contrast to cognitive control.

What Have I Learnt About Mathematics Education from Indigenous Contexts?

Mathematical thinking can always be based on imagination, experience, memory and embodied knowledge. In EHSM, the opposite is often expected. Typical 'problem-solving' tasks expect students to respond with abstract ideas and methods to quasi-contextual questions, and typical difficulties include students not being willing or able to imagine the contexts and reason accordingly. It may be neither the context nor the student that gives rise to these difficulties, but the expectation that abstract methods, rather than ad hoc methods, should be used in realistic contexts.

In the contexts described in this chapter, efforts have been made to bring the whole of what it means to be human into the task of learning school mathematics. On the other hand, to draw a firm line around any Indigenous ideas would be foolish, since similarities among them can be found across the whole planet. Any generalisations that can be made from these studies ought to give insight into *all* contexts for teaching and learning mathematics, including those where EHSM is the accepted cultural norm.

References

Abreu, G. D., & Carraher, D. (1989). The mathematics of Brazilian sugar cane farmers. In C. Keitel, P. Damerow, A. Bishop, & P. Gerdes (Eds.), *Mathematics, Education and Society, Reports and Papers presented in Fifth Day Special Programme at the 6th International Conference on Mathematics Education,* Budapest, 1988. Paris: UNESCO.

Alberta Education. (2006). *Common curriculum framework for K-9 mathematics: Western and northern Canadian protocol.* Edmonton: Alberta Education.

Barton, B., Fairhall, U., & Trinick, T. (1998). Tikanga Reo Tātai: Issues in the development of a Māori mathematics register. *For the Learning of Mathematics, 18*(1), 3–9.

Bishop, A. (1988). Mathematics education in its cultural context. *Educational Studies in Mathematics, 19*(2), 179–191.

Borden, L. (2011). The 'verbification' of mathematics: Using the grammatical structures of Mi'kmaq to support student learning. *For the Learning of Mathematics, 31*(3), 8–13.

Dienes, Z. P. (1969). *Building up mathematics*. London: Hutchinson Educational.
DfE. (2016). *Achievement of 15-year olds in England: PISA 2015 National Report*. https://assets.publishing.service.gov.uk/government/uploads/system/uploads/attachment_data/file/574925/PISA-2015_England_Report.pdf. Accessed 11 Apr 2020.
Doolittle, E. (2006). *Mathematics as medicine*. indigenousmathematics.weebly.com/uploads/2/0/4/5/20451709/medicine_-_doolittle.pdf. Accessed 11 Apr 2020.
Doolittle, E., & Glanfield, F. (2007). Balancing equations and culture: Indigenous educators reflect on mathematics education. *For the Learning of Mathematics, 27*(3), 27–30.
Freire, P. (1972). *Pedagogy of the Oppressed*. (Myra Bergman Ramos, Trans.). London: Penguin.
Fyhn, A., Eira, E., & Sriraman, B. (2011). Perspectives on Sámi mathematics education. *Interchange, 42*(2), 185.
Fyhn, A., Nutti, Y., Nystad, K., Eira, E., & Hœtta, O. (2016). "We had not dared to do that earlier, but now we see that it works": Creating a culturally responsive mathematics exam. *Alternative: An International Journal of Indigenous Peoples, 12*(4), 411–424.
Hunting, R. (1987). Mathematics and Australian aboriginal culture. *For the Learning of Mathematics, 7*(2), 5–10.
Jorgensen, R., Sullivan, P., Grootenboer, P., Niesche, R., Lerman, S., & Boaler, J. (2011). *Maths in the Kimberley: Reforming mathematics education in remote indigenous communities*. Queensland: Griffith Institute for Educational Research.
Kisker, E., Lipka, J., Adams, B., Rickard, A., Andrew-Ihrke, D., Yanez, E., & Millard, A. (2012). The potential of a culturally based supplemental mathematics curriculum to improve the mathematics performance of Alaska Native and other students. *Journal for Research in Mathematics Education, 43*(1), 75–113.
Lipka, J. (1991). Toward a culturally based pedagogy: A case study of one Yup'ik Eskimo teacher. *Anthropology and Education Quarterly, 22*(3), 203–223.
Lipka, J., Hogan, M., Webster, J., Yanez, E., Adams, B., Clark, S., & Lacy, D. (2005a). Math in a cultural context: Two case studies of a successful culturally based math project. *Anthropology & Education Quarterly, 36*, 367–385.
Lipka, J., Sharp, N., Brenner, B., Yanez, E., & Sharp, F. (2005b). The relevance of culturally based curriculum and instruction: The case of Nancy Sharp. *Journal of American Indian Education, 44*(3), 31–54.

Lipka, J., Adams, B., Wong, M., Koester, D., & Francois, K. (2019). Symmetry and measuring: Ways to teach the foundations of mathematics inspired by Yupiaq elders. *Journal of Humanistic Mathematics, 9*(1), 107–157.

Meaney, T., & Evans, D. (2013). What is the responsibility of mathematics education to the indigenous students that it serves? *Educational Studies in Mathematics, 82*(3), 481–496.

Meaney, T., Trinick, T., & Fairhall, U. (2013). One size does not fit all: Achieving equity in Māori mathematics classrooms. *Journal for Research in Mathematics Education, 44*(1), 235–263.

Moss, J., Bruce, C., Caswell, B., Flynn, T., & Hawes, Z. (2016). *Taking shape: Activities to develop geometric and spatial thinking. Grades K-2*. York: Pearson Canada.

Moss, J., Caswell, B., Hawes, Z., & Jones, J. (2017). Prioritizing visual spatial mathematical approaches in first nation early years classrooms. In A. Chronaki (Ed.) *Mathematics Education and Life at Times of Crisis, Proceedings of the Ninth International Mathematics Education and Society Conference*. 2, 728–737.

Perso, T. (2006). Issues concerning the teaching and learning of mathematics and numeracy in Australian schools. *Australian Mathematics Teacher, 62*(1), 20.

Richards, S. (2010). *Improving maths education in Honduras*. New York: Ford Foundation.

Robinson, J., & Winthrop, R. (2016). *Millions learning: Scaling up quality education in developing countries*. Washington, DC: Center for Universal Education at The Brookings Institution.

Sterenberg, G., Barrett, L., Blood, N., Glanfield, F., Borden, L., McDonnell, T., Nicol, C., & Weston, T. (2010). To become wise to the world around us: Multiple perspectives of relating indigenous knowledges and mathematics education. *Vinculum, 2*(1), 7–19.

Trinick, T., Meaney, T., & Fairhall, U. (2017). Cultural and mathematical symmetry in Māori meeting houses (wharenui). In M. Rosa, L. Shirley, M. Gavarrete, & W. Alangui (Eds.), *Ethnomathematics and its diverse approaches for mathematics education* (pp. 235–255). New York: Springer.

Verran, H. (2000). Aboriginal Australian mathematics: Disparate mathematics of land ownership. In H. Selin (Ed.), *Mathematics across cultures* (pp. 289–311). Dordrecht: Springer.

Vithal, R. (2003). *In search of a pedagogy of conflict and dialogue for mathematics education*. Dordrecht: Kluwer.

7

Some Mathematics Education Alternatives in England

Dimensions of Care in Alternative Educational Spaces

As I said at the start of the book, I am asking 'alternative to what?' In the previous chapter, I looked at some contexts that are not only outside mainstream education but also outside the dominant culture of their countries.

The cases in the previous chapter exhibit care for learning mathematics through the importance of care for the community, for mathematics and for learning through relationships. This is manifested through:

Care for Community

- support from outside the classroom, constructing a balance between the values and goals of European Heritage School Mathematics (EHSM) and local sources of knowledge and need;
- entwining EHSM with other sources of knowledge of space and quantity;
- relationships between local methods and conventional mathematical terms;

A. Watson, *Care in Mathematics Education*, Palgrave Studies in Alternative Education,
https://doi.org/10.1007/978-3-030-64114-6_7

- how everyday experiences give purpose and insight into quantification, movement, position and shape;
- the difference between grafting mathematical abstractions *onto* everyday experiences and recognising abstract structure *in* everyday experience.

Care for Mathematics

- using language and dialogue to describe, reason and clarify;
- coordinating multiple meanings and uses of number;
- using of verbs instead of nouns to describe and elaborate concepts;
- the central role of fundamental spatial perceptions, including balance, symmetry, pattern, position and measurement.

Care for Learning Through Relationships

- mutual respect;
- learning how to learn in school;
- whether and how teachers bring themselves into their work, such as being alongside learners;
- tenderness about identity, shame and learners' priorities;
- yarning to reflect on learning;
- the holistic nature of coming to know anything.

All these were in contexts where students came from historically underachieving groups and care was understood to be a necessity in educational systems and projects. I now move homeward to some non-mainstream spaces in England that are surrounded by a dominant national culture of examination-focused education and school as the provision of knowledge, skills and the habits of learning. Care as identified above therefore has to become evident in the educational spaces and institutions since it seems not to be prevalent in the surrounding culture.

My sample of alternative educational spaces in England is opportunistic and is not representational or generalisable. It illustrates what is possible when an educational space is outside the mainstream and can develop its own priorities and norms rather than following perceived national constraints (see Chap. 1). Most, but not all, such spaces are guided by the National Curriulum NC; for some this is so that they can operate in an educational marketplace, for others it is to give students access to nationally recognised measures of success. I found educational spaces that are different from privately funded versions of mainstream norms and practices. I selected those for which there is an identifiable ethic of care for students and mathematics that goes beyond compliance with testing regimes. Given their freedoms from statutory control, how is care for mathematics, care for students and care for learning mathematics operationalised and how do their practices relate to the ideas generated in earlier chapters? The spaces vary from home-schooling, provision for students with characteristics that mean they cannot be educated in mainstream settings, and also an expensive private school operating academic selection. The students in these spaces vary widely from those born into privileged classes who expect to benefit from existing social structures, those whose parents have an alternative view of the aims of education and have the resources to choose, and those for whom some special provision is appropriate. In all of the spaces I have chosen for this chapter a recognised qualification in mathematics is understood to be important as a passport to future study or employment.

Home-Schooling

I turn first to two opportunistic home-school contexts, both concerned that education should develop their children as interesting and interested people. Home-schooling varies widely in aims and quality and can evoke strong reactions from many directions. These two contexts are in no way offered as representative of anything other than educational spaces in which care for the children's learning of mathematics is manifested.

Trina home-schools her only son, Matt, who was 5 ½ when we talked. The medium-term plan is that he might go to school but Trina and her husband wish to avoid formal learning until he is older. They want him to experience the interconnectedness of knowledge that in school is often separated into subjects, and also to understand complex situations first-hand. For example, they have taken him to a desert to learn about desert ecosystems, and regularly attend an adult poetry group so he can learn how people interact with poetry. Trina says: 'We are committed to prioritise the integration of body, spirit, heart and mind for Matt, which in our view cannot be developed away from the parents from 8am till 3pm at this age'. Trina's beliefs about learning are that schooling should follow and nurture children's own interests. They take an inquiry approach to life in general together, following up his questions and proposing new ideas and suggestions, integrating sports and outdoor activities into his life, and interactions with a range of people. When we spoke he was passionate about diggers and building sites. During my interview with Trina he spent time watching and identifying birds through binoculars. It is not unusual for him to spend a couple of hours pursuing something he finds interesting, and he also has friends with whom he can sustain this kind of activity.

For mathematics, Trina does not feel able to provide for Matt and also wants to avoid setting up situations that have any negative effect on their emotional relationship. Shared experience works better. She has learnt from the local home-schooling community that it is typical for children to have tutors for mathematics after the age of about 11 and there are also other providers who are used, such as franchises, but Trina resists provision that offers linear pathways of learning. Plenty of mathematics has already emerged between them. Matt and Trina have compared volumes, lengths and capacities as these arise in everyday activities and special contexts, such as during a visit to a river museum. They have talked informally about ratios when cooking; ideas such as 'more than' and 'less than' and approximation arise naturally in that context. He started learning numbers because they play dice games together, his general knowledge of number grows when they use numbers and quantities in everyday life. They have started to use Lego™ bricks as a source of questions, for example 'How many of this size are needed to make that size?' using numbers in the 2,4,8 family. Learning about time has emerged through interest in analogue clocks.

Trina says:

> I am determined to find a way to do something more meaningful with maths but it means I have to learn or I have to find the person to do it. Perhaps someone could develop a curriculum with pointers and suggestions for people like us or other parents who would like to introduce maths not as numbers but as relationships between things.

Ed educates his own three children at home, having decided to do so because their local school adopted a militaristic line on behaviour management, while his own parenting approach is more concerned with restorative justice. I was interested in the mathematical content of the education process in his family. His engineering background and training as a primary school teacher means that he can provide for mathematics with confidence. Ed's approach is to watch for their interests and provide tasks, materials, prompting questions and exploratory tools to support what they do. They also have access to sequenced online materials and demonstration videos that they dip into when appropriate. He talks in terms of their cognitive development leading their interest, so that being interested in counting arose when they were about 4 and 5, multiplication when they were about 7 and 8, and one of them is now interested in the abstractions of elementary algebra at the age of 12. Counting arose in games, stories, rhymes and daily actions. More obvious examples of his deliberate contribution to their learning is that he can devise tasks that help them think beyond techniques found online. One example he gives is that following an interest in Pythagoras he showed them that any triangle can be split into two right-angled triangles and therefore be subject to the rules of Pythagoras' theorem and will display the trigonometric ratios. A task that revisited these ideas was to estimate the size of the complete ammonite from a fossil fragment in their home (see Fig. 7.1).

The diagram on the right of Fig. 7.1 shows how they represented one sector of the outer ring of their ammonite fragment. They measured its inner and outer arc chords and assumed that a rectangle and a right-angled triangle could be used to estimate the angle at the centre, using Pythagoras, inverse sine and similar triangles. From this, they estimated the radius of the original complete ammonite with Ed leading and questioning and providing much of the symbolisation.

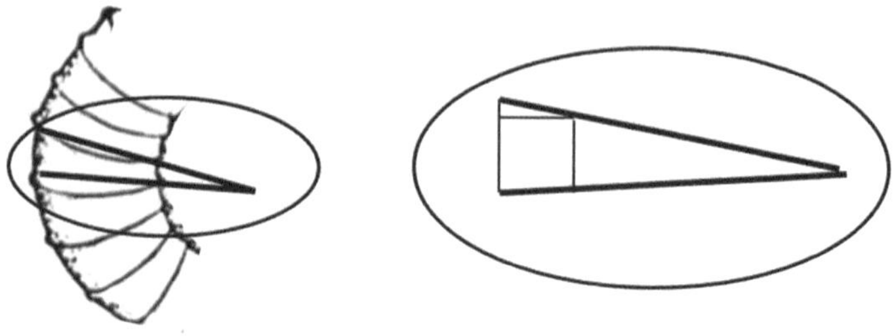

Fig. 7.1 Diagrams for estimating the radius of an ammonite from a fragment

A deliberate move to teach decimals came from his realisation that they had not 'done' them yet. He cut a sausage into ten equal parts and they had a conversation about tenths, hundredths and thousandths (that would have required some imagination). Because they had a strong understanding of place value with whole numbers they were able to extend this into the decimal fractions parts of numbers, and operate arithmetically on them without any conceptual difficulty. He also used soya beans, which are roughly spherical, to introduce a need to measure in millimetres to answer one of the children's questions—how full is the soya bean jar?

Classifications, relationships and properties of shapes arise in art and craft activities, as does some measuring and ideas of similarity and congruence. Writing, diagrams, sketches and workings-out are often done on A1 sheets of paper on which they all write in collaboration.

It is also important that they eat together every evening and talk about the day, so that reflection on what has been thought about, observed, done and achieved is embedded in family language. Ed says: 'Dinner is the best time for learning'. He recognises that, as well as his own educational background, it is important that he has 'a good ear' for listening to children and listens to pick up clues about what they are interested in.

It would be easy to dismiss the care, knowledge and love that goes into his home-schooling as special because they are his own children. However, he told me that a school with which he is associated has recently started developing nurture groups for children who are, for various reasons, troubled or delayed in school and need to get more specialist attention in smaller groups. He believes that his style of teaching, responsive and based on listening, would be valuable for such groups of students.

Important features of his teaching are: it is led by children's interests; he is a knowledgeable teacher; there is reflective conversation every day; he designs imaginative tasks; he, and therefore his children, see the world and all activity with eyes for the potential of mathematising and mathematical talk; his caring is natural; home life is activity focused with shared growing, cooking, playing, making, and so on. Mathematics is a mixture of contextually relevant ideas and attention to the formal, abstract aspects of the subject. However, the relevant mathematics does not stop at what is needed for the immediate task, it can lead to exploration on its own behalf; conversely, formal and abstract ideas do not stop with the formal teaching presented on videos, they are expanded and explored through his design of follow-on tasks. The constraints of examination syllabi are not observed. Despite Trina's claims to be unable to provide for the subject, mathematics with Trina and Matt has many similar qualities to Ed's work with his children. Homelife is rich in the mathematics that arises in shared activities and talk and number-related games. However, Trina lacks the experience and confidence to devise challenging follow-on tasks. As with Ed's children, there is obvious care for the student, and for mathematical learning, but it seems that deeper care for mathematics may have to be found elsewhere.

In these two cases, the didactic model developed in Chap. 6 dissolves into a family model in which love for each other and love of inquiry shape learning. The idea of learner and adult alongside each other in a shared task dominates these educational spaces.

An Independent Therapeutic School (ITS)

Ed's concern for children with fractured backgrounds leads me to write about a residential school for primary school-age children with extreme social, emotional and behavioural needs. The children are all within the usual range of cognitive capability, as far as anyone can know, but their traumatic infancy and childhood may have had a lasting effect on some aspects of brain development. For example, their ability to focus can be affected by the need to be constantly aware of changes in the environment; their relationships with adults might be fearful and characterised by defensiveness or by a superficial need to please. Their environment has led to patchy educational experience so that, as well as not being able to live in a family situation, they have learning delays and also find it hard to fit into social norms. The academic provision in this school takes place in classrooms in a typical primary school setting and follows the National Curriculum. Progress in learning, and in willingness to learn, is one of the aims of ITS, so that children can return to local schools after three years. The first year involves orientation, establishing relationships and an individual programme of support, the second focuses on working with the trauma and the third prepares them for re-entry into family and full-time outside life, including mainstream school. Although the school is independent all the children are funded from public money; independence means the school can maintain its principles and not be subject to political whim.

Because ITS is residential, with support elsewhere for the family, education is aligned with the culture within which the children are living. What takes place in the classroom in terms of pedagogy, learning and knowledge is aligned with life inside and outside school—what Noddings would call the immediate others—and with the cultural expectations that will face the child when they go elsewhere. The collective development of individual programmes is an intense, continuous, version of Lochville's practice of individual student-centred planning with multiple participants (see Chap. 6). Care for emotion is integrated with care for learning.

Children are grouped for schooling according to emotional development rather than by age. Groups therefore include mixed prior knowledge and experience. Because many students have suffered early childhood trauma, pedagogic principles associated with early years up to age 5 by statute are applied throughout the school for all ages, namely:

- A unique child: every child is unique, constantly learning, and can be resilient, capable, confident and self-assured;
- Positive relationships: children learn to be strong and independent through positive relationships;
- Enabling environments: children learn well in enabling environments in which their experiences respond to their individual needs and there is a strong partnership between practitioners and parents/carers;
- Learning and development: children learn and develop at different rates (DfE 2017).

Mathematics is taught with a mixture of formal and informal activity appropriate for the child and the mathematics. Significant differences in the teaching they have here and what they may have had outside are the continuity and regularity of their mathematics experiences that can be tailored to their own conceptual development, and the close personal attention of a trained teacher as well as other supporting adults. A school principle that is also carried through in classrooms is that children are learning to live *as* a group, not only *in* a group, so that they can better cope within the dynamics of life outside school. This drew my attention to differences between students learning *in* a group, *as* a group, *from* a group, or *how to act in* a group. Literature about groupwork in mathematics groupwork often describes listening, turn-taking, communicating, inclusion and respecting other views—how to act in a group. Some sources emphasise the importance of different roles when a task is being undertaken together—teamwork. Others describe a need for everyone to learn the same core idea from a joint task, the group being the milieu for learning. In ITS, all these possibilities take place.

In common with early years practice, mathematical ideas can crop up and be discussed in any context, so much of the teaching might appear to be play-based. Learning to play, and handle the associated risks and uncertainties, as well as the social and emotional aspects, is something most of the children need to develop but formal learning may feel safer for some of them. For some children, working with others is a welcome experience; for others, it is painful and being alone, or working one-to-one with an adult, is preferable. Staffing ratios are very generous so that any child at any moment can be in a one-to-one relation with a caring adult alongside. After the initial stages of settling, the idea is that children can develop new neural pathways and behavioural habits that provide alternative routes to those that have in the past led to failure, eruption of negative behaviour, and even self-hatred.

In the next chapter, some Forest School practices are described, and indeed children in ITS have these too. 'Forest' can be a very scary place requiring courage; for others, the 'forest' is safe as it never rejects them.

The NC in its entirety is too detailed for the timescale of the school and is stripped down to the strands that will make a difference to later re-integration into mainstream: number (including time and money) and some spatial concepts are essential. Children arrive with some memory of the mathematics they have learnt before, if any, and this has to be woven into new learning. Classrooms use a mathematics table laden with concrete materials from which children can choose. Conversations take place in small groups sitting on rugs or round a table. In these talks, teachers can model chains of reasoning and talking about the meaning of symbols, numbers, relations and operations. Recording work is important, and also gives individual children some pride about their thinking. Children at ITS are acutely aware of anything that expects too little, or too much, of them—they have been the guardians of themselves against the world for many years already. There is often some resistance to mental arithmetic rather than written algorithms. This could be about the need to feel safe, or also that what goes on inside your own head can be rather frightening and having something external and physical to do can be more calming. Games and puzzles are helpful because they can embed concepts, provide a time for purposeful practice; they involve nonthreatening cooperation and might even be fun.

Many of the characteristics of care that were evident in Indigenous contexts are also evident in ITS and in the home-schooling practices described so far. A major difference is that something takes the place of the culture of the surrounding community. ITS students are shielded from the outside to some extent but there is support for families going in in their home communities while they are at ITS. The expectation is that they will return to a supported situation. However, EHSM culture is dominant at ITS—the two differences from mainstream are that expectations about prior knowledge are lower because of missed education so far, and that emotional needs have priority.

Rathbridge School

Equality of voice is embedded in the ethos of Rathbridge school, an independent school with boarders and day pupils. It is governed within the ideals of a faith community with Christian roots that has a commitment to equality, truth, peace and community. Its students are from all faiths and none. It has a reputation for regarding every child as precious and equal and provides bursaries for those who have a special need to be there, such as those with autism, dyslexia or mental-health difficulties but can cope, with support, in a school setting. The school is small so that all students are known to all teachers. Relations between teachers and students are relaxed and friendly. Children come into the secondary phase from a range of primary provision.

I saw four secondary mathematics teachers at work, all who had previous experience in mainstream schools, and had conversations after I had observed. The enacted nature of mathematics seemed to be similar to what I see in mainstream schools. It is hard to generalise from brief observations of teachers but mathematics as 'things to be understood so that things can be done' and 'things to be done assuming understanding comes later or came earlier' seemed to be a norm, as it is in many classrooms where the dominant context is a test-driven school mathematics culture. The 'test' context seemed to be accepted as the frame for their work, partly explained by the school's need to survive in a competitive marketplace. Within this frame, then, rather than wishing there were other aims, what can be said about care for mathematics, care for students' cognition, and equality?

The whole school exudes an ethos in which students are listened to, young people are respected as equal to each other and to adults in the right to express a viewpoint and be heard. How does this affect mathematics lessons? At first, I could not see anything obvious, but the teachers told me that: 'I can teach in the way I want to teach' and 'my style fits the school' and also commented on what is possible with smaller class sizes when there is a high level of neurodiversity in all classes. I realised that I was seeing more attention to what students said, more respectful listening, more conversations that positioned teachers and students equally, than I saw in larger classes in many state-funded schools. The conversations were initiated either by students or teachers and usually involved questions about understanding the mathematics, or puzzling about what had gone wrong.

In every class, teachers have to be aware of a range of special needs, sometimes preparing special materials for individuals, sometimes teaching in ways that are necessary for some while benefiting everyone. Classes are small so that individual attention is possible for all. Teachers also willingly give their non-teaching time for drop-in help. They told me that students are respectful of each other and hence are not reluctant to say when they do not understand something, or when they need help. A balance between getting through the curriculum in a given time and working on deeper understanding feels less pressured that in the teachers' earlier mainstream experience. One teacher said that in a mainstream school she felt she had been expected to 'teach the quickest way' all the time, such as memorising solution methods early in her teaching of trigonometry rather than working on understanding the ratios. Here, especially within its emphasis on outdoor education, they could approach trigonometry by spending time finding heights of trees and also, with the need for many students to visualise and/or engage with something active, they could draw triangles, measure and calculate. It was sad that they had felt unable to do this in their previous mainstream settings. I noticed the same kind of monitoring of progress and use of data to predict grades that is predominant in the public sector, so why would they feel less pressured? Maybe this is because of small class sizes, or students' willingness to ask questions, or because teachers use a variety of approaches taking account of students' needs, or because the school is a calm, quiet and friendly place, or a combination of all of these. Maybe it is because of the successful and caring integration of students who elsewhere might be kicking against the system. Maybe it is that this school has attracted teachers who have always wanted to work on developing understanding because of the ethos of inquiry and truth, so 'understanding' is the school mathematics culture. The faith group in which it sits has a commitment to seeking truth, so I would expect to see this enacted somehow in mathematics lessons when the subject itself, as school level, is often presented as given, external, truths. 'Understanding' is a loose word open to a range of interpretation so I shall illustrate the meanings I saw in Rathbridge School with some examples.

Set Theory Lesson

The lesson started with students saying what was meant by the symbols ∪ and ∩ in $A \cup B$ and $A \cap B$. Then they were given on cards 16 Venn diagrams and 16 expressions and asked to match them (see Fig. 7.2).

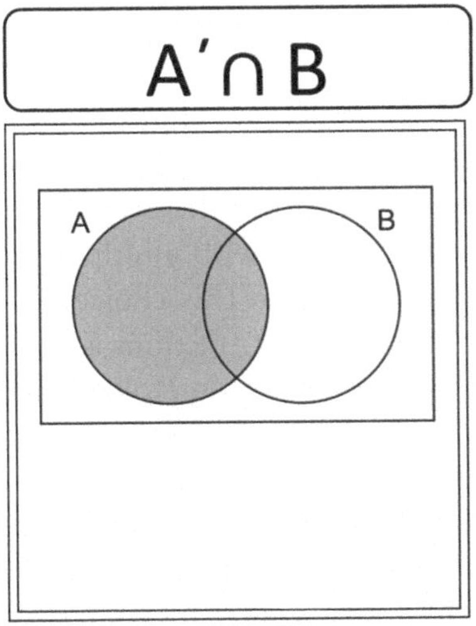

Fig. 7.2 Label and diagram for a card-matching activity for Venn diagrams (copyright©Mathsbox.Ltd)

This involved some cutting and sticking and indeed this teacher (Mary) said she used a lot of 'cutting and sticking' tasks, avoiding unnecessary writing. The students were industrious—it was Friday afternoon and a task with physical materials seemed a good decision so that students had objects to 'fiddle' with as well as quite abstract ideas to get their heads round. Mary had time to talk to each student during this task. Sometimes the student stated their reasoning out loud; sometimes Mary modelled the reasoning by indicating a region and asking what it meant, or by asking the student to indicate a region that represented a particular symbol.

After the individual work the whole class discussed the final matching. Verbal reasoning models thinking and also conversely the structure of speech can itself generate thought (Vygotsky 1987). The talk involved in reasoning about sets was sometimes between people, or one way, or internal. Several students needed to mutter to themselves, and also to indicate inclusion and exclusion by some hand gestures, even when working on their own.

Fraction Calculation Revision Lesson

It would be unfair to take Paula's revision lesson on fractions to be an example of normal practice, but nevertheless there were indications of an underlying care for cognition within the overall caring ethos. Paula had taken the unusual decision to start with multiplication and division, and then do addition and subtraction. This choice is conceptually sensible because students will have first met fractions as 'fractions of …' quantities, from which multiplicative reasoning follows and provides a foundation for proportional reasoning. Most use of fractions within school mathematics is multiplicative; it is in calculating probability and some algebraic contexts that addition of fractions and rational expressions becomes mathematically necessary. Cognitive care requires conceptual coherence, as the multiplicative work displays, but also requires the development of familiarity with imposed testing procedures, justifying the additive work.

Paula's commitment to developing understanding was clear even in this whistle-stop tour of fraction calculations. On the board was a rectangle diagram (see Fig. 7.3) to show $\frac{1}{4} \times \frac{2}{3}$ which had also been written as '$\frac{1}{4}$ of $\frac{2}{3}$' since this makes sense of $\frac{1}{4}$ as an operator.

The rectangle diagram shows both fractions as operators, with first $\frac{2}{3}$ being shaded and then a quarter of that being further shaded. The rectangle becomes 12 units of $\frac{1}{12}$ by this process, and only $\frac{1}{6}$ ends up being double-shaded. Such a model can be used to show that 'multiplying tops and bottoms and simplifying' works, also to infer a numerical method from the shading, also to retain the 'operator' meaning of both fractions on 'the whole' that is the rectangle. They then moved on to division. Since this was a revision lesson with students of multiple backgrounds she asked them to explain how to do: $\frac{3}{4} \div \frac{1}{8}$.

Student 1: Is it three times eight divided by four times one?

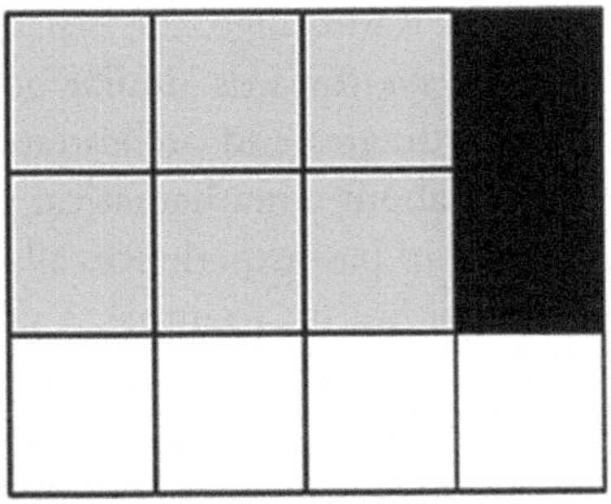

Fig. 7.3 Fractions multiplication displayed as a rectangular array

Student 2: is it turn three quarters into four over three and multiply?
Student 3: is it keep, flip, change?
Paula: Oh, you have heard of 'keep, flip, change'

She did not continue to use this language herself but wrote: $\frac{3}{4} \times \frac{8}{1} = \frac{24}{4} = 6$.

She pointed out that cross-multiplying would also work, indicating the method given by student 1, and asking him 'which goes where?' since there are two possible orders for cross-multiplying, only one of which retains the meaning. After this, she drew a diagram 'to show *why* it works' in which she indicated the number of times that one eighth 'went into' three quarters. This would be a concept for division they had met in primary school, for example: 12 ÷ 3 could be interpreted as 'how many threes in twelve?' rather than 'split twelve into three equal parts'. So, even in this revision session, an underlying meaning of fractions as operators on 'wholes' is presented visually to remind students that these calculations are not merely about performing methods.

Shared principles among teachers do not necessarily lead to similar pedagogy since different teachers interpret them into action differently. In part, this is to do with knowledge of what different forms of pedagogy are possible but it is also to do with the autonomy of teachers. When I watched a lesson taught by Kathy, the head of department, I took lots of notes about what I saw but it was through conversation afterwards that I understood her principles. In Rathbridge, the belief is that different students follow different pathways towards similar goals, possibly understood differently. Kathy coordinates and orchestrates students' different ways of seeing and thinking about a mathematical problem. They come into her classes with a variety of past experiences all of which can be useful for the learning of others. So, for example, I saw mini-whiteboards used for students to display their thinking as they solved for x in the equation: $3x + 6 = 24$, the second of a sequence of three equations. She asked them explicitly to look at each other's method. One student said, 'All my working is there but I write it randomly', but most had used a coherent downward layout, some writing each step explicitly ($- 6$ and $\div 3$, or $\div 3$ and -2) while others wrote each successive transformation only. Kathy had asked them to solve a sequence of equations: $3x = 24$; $3x + 6 = 24$; $3x - 6 = 24$ so that they could see the difference between each solution method. The next example after this sequence was $5(x + 2) = 35$ and she said, 'There are two options', and both were tried—expanding the brackets first before further action and dividing both sides by 5 first to simplify the equation. There was never a sense of being given only one way to do things. Kathy says she cannot teach 'my way is best' because she wants to be more a facilitator (her word) thus fitting with the school ethos of valuing students' viewpoints. She said: 'Here it is alright to be different and those who find it hard or impossible to "fit in" can do well at this school'. It is the norm to care about learning and each other, and this makes it possible to teach in an inclusive way. This autonomy of learning was illustrated in a lesson with another teacher when one student announced to the whole room 'I need to be more careful about my sketches and what they show' while working on some function transformations. The search for truth in mathematics lessons is a search for conventional truth through guided understanding along individual pathways. Just as in John's lessons below, there is nothing cosy. She says: 'I want to get them to do more mathematical work than I do'.

Kathy reports that some students love the abstract nature of relationships within mathematics; others seem to prefer to work with materials; others want a 'realistic' context for mathematical ideas. She used to think this was a result of their past experiences of doing mathematics, but now thinks there is a personal element to these preferences too and her teaching needs to reflect this. For example, she could use trigonometry to solve some snooker problems, or snooker to motivate trigonometry, and those who prefer an abstract approach will get that via diagrams or mental imagery. Occasionally she finds a student whose ways of thinking are a mystery. Here she has time to think about this; in a class of 30, she remembers, this personal attention was very limited.

John: Secondary Mathematics

John has taught mathematics and held leadership positions in several schools, mainly independent, some selective and some not, the latter schools including students with language, emotional and other difficulties. A few years ago, John was teaching A-level mathematics in an ancient selective high achieving school and introduced ideas he had found through reading about the Harkness method used at Philips Exeter Academy, a high achieving US boarding school. This method is named after a philanthropist who introduced the idea and depends on having small classes (John thinks the maximum would be 12) and is seen as an alternative to a teacher 'lecturing' from the front. Lessons are always discussions between everyone present arranged round an oval table with no hierarchical positioning and without the teacher playing a pivotal role. As everyone can make eye contact with everyone else discussion can take place directly between students instead of through the teacher as moderator which would be more usual. John appoints a 'chair' for the lesson whose role is to orchestrate inclusive discussion so that everyone can speak or keep quiet equally, including the teacher. John found that the amount of his own talk reduced over time as he became more confident with the method. What was there to discuss? Students prepare for a lesson by working on questions John has posed beforehand, so they all have something to bring to class for discussion. When he first introduced this way of teaching, some students did not do the preparation since it was not structured like their familiar homework, that is problems using ideas from the previous lesson. When they realised that being unprepared meant they could not participate they changed their habits. Designing homework tasks was a major difficulty for John and continues to be so when he uses similar teacher methods with other students in other schools. John wants discussions to be about mathematical concepts rather than methods so needed to ask questions that were not closed and required thought rather than, say, Internet searching or imitating worked examples. He describes his role as 'earning my money outside the classroom' in question design and one of his measures of success is that students have commented that he makes them work 'too hard'. It has also always been the case that some teachers set homework that is exploratory, or about gathering data, or some other preparation for future lessons. John's method is alternative in that the discussion in lessons is not further teaching led by the teacher but is student-led discussion about their preparatory experiences; the homework is to explore mathematical issues. An early question that was successful in terms of promoting mathematical discussion was to explore logarithms and why the rules of using them work. This is a typical sequence of preparatory prompts that guide the process:

1. Write $\sqrt{2}$ as a power of 2. Hence, express $4\sqrt{2}$ as a power of 2. Hence solve this equation:

$$2^{(3x+4)} = 4\sqrt{2}$$

2. Write each of the following as a power of 2: $\sqrt{2}$; 8^x. Hence, solve this equation:

$$8^x \times 2^{(x+1)} = \sqrt{2}$$

3. Given the following facts:

$$\log_{10} 10 = 1; \log_{10} 100 = 2; \ \log_{10} 1000 = 3; \ \log_{10} 0.1 = -1$$

Can you form a sentence that replaces the word 'log' in the given facts? What would $\log_2 4$ be?; $\log_3 81$?; $\log_8 2$?

On moving to a school with larger classes he extended his question bank to include the mathematics being learnt by younger students, for example 'what does sine 45° mean?' and 'how does addition relate to factorising?'. With larger classes he could not provide a round table but lessons still involved discussion, student presentations and peer-tutoring. His own questioning developed and he had standard questions such as 'why do you think that?' and 'where does that come from?' and 'discuss that point with each other' as well as topic-specific ones. He reports one student in a year 7 class who was 'maths-phobic' (his word) ending up as one of the highest achievers in the class. This, he thinks, is because the environment was emotionally safe; they could ask 'why'? without fear. This student was the first to offer a conjecture about the possible meaning of $a^{1/2}$ based on the realisation that $a^{1/2} \times a^{1/2}$ had to equal a^1 and therefore had to be the square root of a. The rules were not arbitrary, they had mathematical meaning; authority had shifted from the teacher to the mathematics. His core belief about students is that 'they like to talk; to interact; and to have their opinions heard' and that to learn well 'you have to think about it; you cannot be given everything'.

I asked him whether he was assuming that *all* students could engage in reasoning through talk. He gave me the example of a secondary state-maintained school he had worked in for which the demographic was a mixture of well-resourced students from educated backgrounds and students who were disadvantaged and from typically underachieving backgrounds. Those from challenging backgrounds thrived in the atmosphere of discussion, of being supported through talk and the focus on making meaning rather than testing. Those from higher achieving backgrounds tended to be more test-focused and wanted to be given methods rather than challenged to think; his method worked best for those with fractured past experience. Teacher–student relationships in which anyone can challenge anyone are egalitarian.

His practice is in contrast to normal mathematics teaching in most secondary schools and has not always been in accordance with normal practice in the schools where he has taught; there has to be a period of development of trust especially when students' (and possibly parents') expectations are strongly in another direction. The foci on mathematical reasoning and equality of voice are unusual and it is striking that this approach has worked with a range of students, not only from high-achieving, vocally confident backgrounds in small classes but also with socially mixed groups in larger classes.

Reflection

The educational spaces described in this chapter have differed widely, catering for the most to least privileged in society but all having features that are alternative to the surrounding norms of their respective communities. John's teaching methods are not normal in expensive independent schools; Rathbridge is demographically different to most independent schools that have boarding and day pupils; Ed and Trina's children are removed from the schools used by most; ITS is screened from all outside communities. Yet they have many similarities in their care for the students and the learning of mathematics.

All depend on trusting relationships and listening skills between teacher and students. All regard students as potential or actual thinkers in the classroom. All care for mathematics cognition by giving students a voice, either for mathematics or for other aspects of life that bother them or both.

Small teaching groups allow for individual students' ideas to be heard and to shape what happens in a lesson. This means the authority for mathematics shifts from a teacher or textbook to the conversation between adult and student and even to mathematics itself. In John, Ed and Trina design tasks and then play an equal part in conversations. In the other cases, the design and provision of resources and the responsive guidance of conversations and activity are the adults' responsibility. Use of materials and the outside is a feature of most of these contexts, and the role materials play in learning mathematics can be to support attention as well as to model, or act as a source for, mathematical ideas. Dialogue is understood as important in all these cases, not for transference of technical terms, methods and definitions but for developing knowledge through interaction. Adults and students are often working alongside on mathematical tasks. In all cases, the care for students reflects emotional development, and from the early years' perspective in ITS to the ceding of control in John's lessons there is mutual respect. This connection is two-way; learning mathematics supports emotional development and provides a context for healthy relations with adults. The adult's care for mathematics matters too; in most cases they are well-qualified to fulfil this role. None of these are 'anything goes' educational spaces. Ed matches what happens to theories of cognitive development; Trina looks ahead to what might be expected when Matt goes to school; Rathbridge fits with public norms because of marketplace pressures.

There are parallels with practices in the previous chapter:

- the absence of 'talking at' and the primacy of 'listening to'; response to students' ideas, language and sense-making; adapting teaching to support learning;
- working with what students bring into the educational space; incorporating students' emotional and social selves in mathematics learning; treating students as individuals within wider society and communities;
- design and choice of tasks that support meaning-making; using multiple materials and representations; using 'outside' as a source for mathematising;
- expecting and supporting capabilities for abstract reasoning; teachers and students working side by side on mathematical tasks;
- curriculum norms being respected, but students' interests and strengths often take priority.

In the next chapter, I present examples from within mainstream publicly funded mathematics teaching where individual teachers imbue their mathematics teaching with similar care.

References

DfE. (2017). *Statutory framework for the early years foundation stage*. https://assets.publishing.service.gov.uk/government/uploads/system/uploads/attachment_data/file/596629/EYFS_STATUTORY_FRAMEWORK_2017.pdf. Accessed 11 Apr 2020.

Vygotsky, L. (1987). Thinking and speech. In R. W. Rieber & A. S. Carton (Eds.), *The collected works of L.S. Vygotsky, Volume 1: Problems of general psychology* (N. Minick, Trans.). New York: Plenum.

8

Care in Publicly Funded Mathematics Education

Introduction to Care in Maintained Schools

Nothing of what I have written about care has much point for me unless it is possible to exercise similar care in the educational spaces that are the norm for most school students in England with the constraints of the national curriculum (NC) and testing regimes. As I said in Chap. 5, I believe that all teachers care. However, there are special features of care for students' learning of mathematics that have been identified in Chaps. 6 and 7 that go beyond merely adding 'care for students' to 'care for mathematics', or vice versa.

In this chapter, I illustrate how some teachers can exhibit care for mathematics, care for learners and care for their learning of mathematics within mainstream educational spaces in England. I start with some for which additional funding or internal management have provided smaller classes than usual in recognition of need. Then I move to classes that are the usual size of around 30 students. I aim to show that the caring qualities for students, mathematics, and the learning of mathematics seen in some alternative spaces are possible within mainstream spaces.

A. Watson, *Care in Mathematics Education*, Palgrave Studies in Alternative Education,
https://doi.org/10.1007/978-3-030-64114-6_8

The cases I use provide evidence of successful practices, because each of these cases has a record of high, and sometimes unexpected, student progress and attainment. Some also provide evidence for social and emotional progress as well. These are 'alternative' in several ways. Perhaps lessons are structured unlike most mathematics lessons in some way. Maybe the articulated and demonstrated priorities of the teacher are not in precise accordance with the norms of the NC, such as the year-by-year expectations for primary schools, or the examination goals for secondary schools. Many schools and multi-academy trusts expect lessons to follow particular formats, textbooks, lesson plans or slide shows. The teachers in this chapter generally do not follow these strictures, or when they do it is moulded within more complex personal intentions as well as imposition from outside. They all aim to respond to students, not giving 'coverage' priority over learning.

A current focus on mastery has, in many schools, become a one-size-fits-all frogmarch through the curriculum rather than, as mastery was originally conceived, an expectation that everyone will understand a core idea before the class moves on, with extra support for those who need more teaching. Cuts in school funding have limited out-of-class provision and counselling that might enable all to access the curriculum. In practice, these might only be provided by teachers working more hours on top of normal timetables. The autonomy given to, or taken by, the teachers in this chapter is not available to all because of such restrictions, but there are tweaks that can be made to practice that might make a difference.

In all the cases in this chapter, there is obvious care for individual students and their learning. Every case balances getting through tests, understanding mathematics as a human activity, and care for individual students. These teachers are not offered as models to follow because they work within particular circumstances, but as examples to demonstrate possibilities. These teachers do much more than prepare students for the next test; in general, they do not reduce mathematics to a sequence of things to do and memorise. Several provide counterexamples to the practice of offering low-level tasks and minimising difficulty. None of the teachers ignore the personal lives of their students. All see their task as enhancing the development of reasoning and inquiry in and through mathematics alongside the growth of students' mathematical repertoire. All value the autonomy of students and also exercise their own to various extents.

My reports of practice reflect the ways in which the teachers talk about practice. Often I use their own phrases. The different weightings given to aspects of their teaching reflect their responses to questions and observations, where these were possible. Their teaching demonstrates fundamental care for learners as people who are in the process of becoming well-functioning, knowledgeable adults with useful competence and confidence in mathematics.

I start with some special contexts in mainstream schools, made possible in the way institutions manage their finances, as examples of institutional care for those who are not making the expected progress in mathematics. These examples answer a question posed at the beginning of the book: yes, it is possible to 'upskill' students' basic mathematics, and to develop young minds to exercise their intellect in mathematics.

Out-of-Class Provision

In the years leading up to 2010, there was a research study undertaken to show the effects of a government-funded intervention project (DfE 2011). The project provided 30-minute sessions, out of class, 3 or 4 times a week for a whole term, led by trained adults, on the number knowledge and attitude to mathematics of young children whose achievements so far were in the lowest 5%. It was found that their progress in mainstream classes could be significantly altered by this care for individual learning. The project was cost effective, particularly when compared to the possible ongoing educational and social costs of not attending to early learning needs. The whole effect on the child of being helped to bridge their learning back into mainstream is incalculable. As in DISTAR (see Chap. 4), carers and parents were involved where possible. Each teaching episode had a clear focus built on previous knowledge, the use of concrete and visual materials, probing questions, discussion, identification of patterns, and a choice of opportunities to retrieve, practise and reason with earlier ideas. The method depended on listening to the difficulties experienced by individual students, identified by giving them several possible ways to display understanding. Teachers needed to be flexible and resourceful and to be alongside students during the session, probing, listening and responding and hence displaying their care for that student's learning. The financial outlay was considerable but could be offset against a reduction in the need for special help and support later in mathematics and possibly in school more generally. The funding was withdrawn by a new government in 2010, but the project continued on a school-funded, not for profit, basis in some schools for several years until schools were no longer able to find the money.

Emily is employed to lead the same approach in a primary school in addition to other supportive roles for mathematics. The school budget allows for this because it is seen as so important to enable children with limiting backgrounds to fill gaps in their knowledge, learning habits and competence so they can work on the same mathematical ideas equally with others. The school is in a deprived area of a major English city and about 30% of the children need extra educational support. Her work with the selected students, mainly from years 1 and 2 and older children if a need is identified, lasts for one term and she works mainly with groups of 2 or 3, occasionally one-to-one where this is beneficial. They continue also to be in mainstream mathematics lessons alongside their peers and she keeps in touch with them after her direct work has finished. Children are provided with a pack of manipulatives and some activities to take home; parents and carers are supported to use mathematical materials, models and images through workshops and other interactions.

As with many teachers in this book, Emily talks enthusiastically about her own learning. Working alongside students in small groups, using diagnostic materials and being able to watch and listen closely to their thinking, gives her significant insights into their difficulties and conceptualisations. Her ever-growing knowledge of what has been missed in their learning contributes to the work of all teachers in the school, especially for new teachers. Examples of what she has learnt by listening include confusion about writing 14 when a child thought the '4' must come first in 14 because you say it first; an answer to the question 'how many hours in a day?' was 'it depends what time you get up'; the importance of using real coins to learn about money otherwise some children think that faded coins are different from shiny ones. From these interactions she learns the important detail in mathematical talk and symbolism. Developing mathematical competence is a focus across the whole school and there is an atmosphere of enjoyment around the subject, with above-average results.

When fully funded from public funds there were thousands of teachers available to do this kind of work. This no longer applies but Emily is one of the few fortunate to work in a school that invests in early extra provision. The school mission is to lay foundations for life, and this includes all children being fully integrated into the life and curriculum of the school.

Secondary School Nurture Class

Some secondary schools fund nurture groups for students who, for a variety of reasons, cannot learn in normal classrooms, have had erratic experience of school and are in danger of leaving full time education with few useful qualifications. Some students are often excluded for being disruptive or decide to avoid attending school themselves. Others have mental health difficulties that interfere with attention to mathematics or have developed high anxiety in the subject. Nurture groups are so-called because the students are identified as having emotional and social needs that respond to care while also needing the same kinds of educational aims as their peers. Janet teaches nurture groups for mathematics in a secondary school. She has extensive mainstream teaching experience. In some schools her students would be off-rolled (see Chap. 1) as their emotional and behavioural needs could get in the way of other students' learning, but her students mostly attain high grades in mathematics. The school is financially committed to inclusion.

When students first come to her lessons they know, from her reputation, that they might reasonably expect to get a high grade if they work hard in her lessons. However, they also know that mathematics has been a minefield for them so far, that they have failed more often than succeeded, and that school expectations are generally way beyond what they achieve in terms of behaviour, effort, achievement and relationships with adults. They often bring with them into the classroom a history of distrust, anger, violence, non-compliance, withdrawal, antisocial behaviour, and of adults giving up on them in or out of school. There are many similarities with those who go to ITS in the previous chapter. Janet has three and a half terms to get them through GCSE, with the help of a teaching assistant. The contract is explicit—you work hard and you get your qualification.

Janet breaks school norms in a number of ways. Her classroom is very small, with tables grouped to seat up to 15 students in groups very close to each other.[1] Each student always sits in the same place, and they know who else uses that same space in the other groups she teaches. This gives a sense of belonging, a cosy place that is theirs among the hurly-burly of secondary school life that, in mainstream, usually involves a lot of moving around. The group decide on the level of lighting, and often choose a very low level even when this makes it hard for some of them to read. Anyone who wants to enter the room has to knock and wait for a student to let them in. Every student has a box containing their mathematics kit and a plastic folder in which their written work, artefacts, pen, ruler, worksheets and exercise book are kept. Tensions around arriving at a mathematics lesson without suitable equipment do not arise.

Students who initiate, or get involved in, disruptive events are not excluded. She allows the group to move off-task if some event initiates a need for it and they are often the ones who decide when to resume work. Learning that anger and moods, including her own, are part of being human, not a source of shame, is part of her classroom climate. They know she cares about them, and that the school, by putting them in her group, also cares about them. She knows they want to learn and belong and that previous failure in both aspects contributes to their problems in mainstream classes. They want to put right their alienation but do not know how to do it. They want to learn mathematics because that is what their peers are doing; this gives Janet her starting point.

Rather than revisiting basic number work that has been a site for repeated failure in the past, she starts with something that is generally understood to be a challenge for most students as they move towards

[1] During Covid-19 crisis this had to be adapted.

GCSE—this could be algebra or trigonometry. Her thinking is that this will break a pattern of low self-esteem they have in relation to mathematics, that is that they believe they can only do easy things and they belittle anything they can achieve by calling it 'easy'. Because they have a limited time to learn she focuses on what is needed to pass GCSE, seeing this as her social responsibility. This matches their need for structure, certainty and routines. Focusing on deep conceptual understanding is no longer appropriate, she thinks, and they have already accepted the abstract nature of school knowledge and the divorce of school from everyday life. They never ask 'when am I going to use this?'; they need the grade as a social and cultural passport and it is part of Janet's care for them and their learning that they get it. With more time and mainstream classes she does not take these limited views.

They start, for example, with a syntactic approach to algebraic expressions. She avoids x, and also a and b (to avoid past associations with apples and bananas) and lets them choose letters for variables. Then they move to linear equations that can be solved by sight and agree that these are too easy, so she moves to equations that need an understanding of balance and equality and a method. I asked her if she follows an 'I do, we do, you do' pattern in her procedural lessons but she prefers to think of 'whatever they do, I do'. By this she means she works through questions and methods in real time on a whiteboard, modelling the realities of doing mathematics, that is making mathematical and spelling mistakes, getting stuck, using mental methods to reconstruct forgotten number facts, and maybe running out of space. This gives them a model of mathematical working that they can achieve. They know she has a mathematics degree so it is reassuring for them that working in messy ways is acceptable. However, they also depend on her to be professional about her teaching; nothing is accidental and she is always ready, organised, and has planned for their progress. She gives group choices about the order of topics, not individual choices. She says that individual choice can be bewildering and can augment insecurity. The time for exercising informed choice in learning mathematics has gone.

Another starting point is trigonometry. For this she talks about the Greeks and their pursuit of knowledge so that they understand why Greek letters are sometimes used in questions. It is important that this should look and feel different to anything they have done, and failed at, before. The first lesson sets the scene by labelling triangles, angles and sides and the three words: sine, cosine and tangent. She avoids the word 'ratio' as it is likely to trigger negative reactions from the past. She takes a procedural approach to calculating unknown sides, followed by the use of the word 'inverse' that is probably new for them. A final lesson involves seeing and solving right-angled triangles in, for example, kites or realistic contexts with a focus on the empowerment that comes with knowing a method.

Arithmetic is the mathematical domain that has been most problematic for them. Mainly they have become unable to do anything that has been tested at speed and they believe that it always has to be 'fast'. Once they trust both her and the learning environment, and themselves, a little more, Janet works with them to reconstruct from what they know, patching together 'number' from scratch always including decimals and fractions in the story.

I told Janet how ITS, working with younger children who have similar difficulties, uses early years principles: uniqueness, child-centredness, positive relationships (see Chap. 7). Relationships are important in Janet's teaching; they have to be real, attentive, humane but their negative aspects do not have to be avoided or masked. She says that the uniqueness of individuals is not so appropriate for adolescents since the centre of their orientation is the peer group and her classes tend to operate as one group, even if she tries to split them into two teams for a game or puzzle. Child-centredness is not appropriate—they want her to be in charge of the mathematics. I asked about whether she uses silence. Janet said this would not be appropriate; silence would create too much mental space for negative memories and upsetting ideas to take hold. For this reason, she does not simplify her explanations, nor give instructions in a minimal fashion—she loads their attention with a lot of words and ideas and things to do so that they do not have any gaps through which difficulties in their outside lives can creep in.

I wondered how, among the procedural and mnemonic preparations she makes for GCSE, they are prepared to work on unfamiliar problems where they may feel less secure. She often uses what she calls 'solution

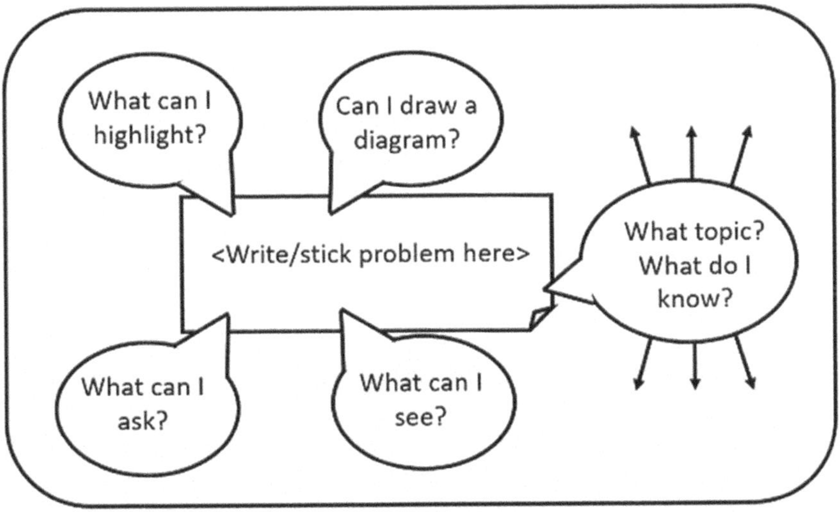

Fig. 8.1 A solution mat to support thinking about unfamiliar problems

mats' derived from her interpretations of the work of Lev Vygotsky and Jerome Bruner (Fig. 8.1 shows a similar design).

The mats, which are for writing on and wiping off, support shifts between representations and abstractions, and between the known and the unknown, when given a situation. Students follow the prompts and make rough jottings on the mats before organising an approach to the question. In her teaching, the actual question posed by authors and examiners might be given after some initial discussion and mat-jotting about a situation. Students also construct their own problem contexts for mathematical procedures, and this prepares them for those the examiners create. In examinations, they often sketch these mats for themselves to organise their thinking.

Because many of her students have suffered some trauma she recognises they need to be mentally strong to deal with some of the contexts used in worded problems. For example, the phrase 'angle of depression' might need strength, and contexts such as rivers and cliffs, common sites for suicide attempts, might also need strength. Financial contexts about interest rates and assumptions about employment in outside life can cause fear.

It is important to restate that this approach is not about preparing people for further mathematical study, or conceptual depth, although these might happen for some students. It is about rescuing their mathematics grade by caring for them, caring for their ways of working and being, and caring for the role of mathematics in their lives, the positive feedback they get from doing it, and the feeling that mathematics plays a part in their skill set. Their social and emotional development, and their self-identity as learners, grow in this albeit test-focused environment.

An Examination Retake Group

The first thing I noticed about Mark was that he was standing at the side of the classroom looking in the same direction as his students at the mathematics on the board, listening to what they say and watching the discussion among the class. He was teaching GCSE resit students in a 6th form college in classes of up to 20. Mainly these students have seen the mathematics in the curriculum before but been unable to pass the exam, so his work is to some extent 'teaching to the test'. However, what he tries to do is more complex than question spotting and procedural practice. His sense of care for students as a component of his work has grown as a result of teaching them and other 'struggling' groups. He has a broad belief in capability that he expresses as 'they *don't* get it but they *can* get it' and sees his job to move them across that gap as much by building confidence and strategies as by building up knowledge. In that sense, he has similar aims to those of Janet (above) but without the extreme psychological and emotional difficulties. He has established a classroom that is friendly and thoughtful and very focused on what to do when faced with mathematics questions, particularly any for which ways in are not immediately obvious. This is as much about their self-belief as their mathematical experience, so he guides them in a variety of ways, sometimes telling and showing, and sometimes structuring task sequences so that they experience early success and then arrive at useful approaches for themselves. He describes his planning as his own battle between open lessons and structured closed lessons. The following task sequence about circles started with open discussion and then moved through a tightly structured section while they prepared appropriate tools:

1. They started with a chat about what they already know about π—when it is used, the calculator display of it, the use of the symbol. He decided not to bother about rounding at this stage as that could introduce an extra problem.
2. Next, he gave substitution questions to position the use of π in algebraic terms, sequencing them towards the area formula: kx, kx^2, to πr^2, and also the formula for circumference. During these tasks, he walked round to check that everyone could substitute before moving on to using the formula. This freed all the students to think about how to use its output rather than still be worrying about substitution.
3. Next, he posed questions involving finding the difference between the areas of rectangles and area of a circle inside it—typical 'shaded area' questions in GCSE. He drew a rectangle to represent a garden and a student suggested a jacuzzi could be a circle inside it. The rest was to be grass. Given that they knew how to find areas of rectangles, and had successfully calculated areas of circles, what could they do?
4. Noting the trust he had in their reasoning capabilities, and that previous failure might be more about confidence and fluency of knowledge, the class then discussed what to do. Once this was established he gave them similar questions with different numbers so that they could connect their 'realistic' question about grass to the abstract 'shaded area' problems, including questions with two circles. Somewhere in this sequence accuracy was introduced with the symbol π itself being understood to be more accurate than a calculator display.

He believes that understanding and confidence are inseparable. His approach to circle area has changed from the previous year when he had used dynamic geometry to give measurements of circumference and diameter that then had to be divided—this gave them a value for π but did little towards their problem-solving confidence and capability.

Care for their learning is not only shown in the detail of his planning and medium-term thinking, but also in his physical presence in the classroom. He makes sustained eye contact; he might kneel or pull up a chair to avoid 'standing over' students and to maintain a shared view of the mathematics; he listens intently and encourages them to do the same. They are sometimes asked to restate in their own words what he and others have said, not to check their attention but to 'hear their own voice'.

He wants them to be aware of their own progress so uses a bank of diagnostic questions as a resource for comparing 'before and after' thinking. An example of this method of working is the question: 'Which multiplier would give an increase of 9%?' with four possible answers offered. Most chose 0.09. Rather than ignore the wrong choice and plough forward with some direct teaching he asked them to think about an electricity bill that goes up by 9% In this context, a student said that 109% was what was wanted, so the multiplier would be 1.09. Mark thought about whether saying '1 plus 9%' or '100% plus 0.09' or '109 over 100' would be good ways to describe this. This level of detailed mathematical thinking characterises his teaching. Giving intense attention to the detail of the mathematics and to students' thinking, particularly those who have had negative experiences in the past, can be exhausting. Mark listens *to* students rather than listens *for* particular answers and this entails significant reflection and replanning between lessons. Because this level of attention is tiring, Mark does not appear to do as some other teachers, namely setting and checking homework and spending time with students outside the class. For him, the lesson is at the heart of learning and that is why he calls his classroom 'thoughtful'. Even with this group that for some teachers might be seen as a 'get them through GCSE' task, he wants them to have a good experience of mathematics. He wants them to feel that mathematics is for them, and has a good name with them, even though for many it has previously been a stick with which they feel they have been beaten.

Both Mark and Janet prepare students for GCSE, and both want mathematics to be somehow part of students' lives. Although they describe their priorities differently, both plan in detail for their students to have the tools to solve unfamiliar questions rather than take a 'question spotting' approach to examinations, so even in these classes there is a combination of 'upskilling' and development of independent mathematical thinking. Both give a level of care for individual learning that is challenging for them.

I now move to some teaching situations that embrace the normal range of attainment in their schools.

Forest School

St Saviours is an urban primary school on the edge of a large town. The teaching I describe is embedded in a school that openly declares values and virtues. During the term when I visited the public declaration was 'Let all that is done be done with love'. Although the school is socially mixed, many of the pupils live in poor-quality high-rise spaces, many parents and carers do menial jobs such as night-time cleaning, possibly having several jobs to make ends meet, and many homes are 'low talk' either because adults are out working or because talk is restricted to information and imperatives. Additionally, many children would have few outside school experiences from which the formalisations of school mathematics could be inductively learnt. For some children, school may be the only place where they are immersed in English. It has a corner of a copse and some scrubland where Sandy teaches Forest School with all the youngest children, taken in small groups. The school sees these sessions as essential for their social and intellectual development, but accountability concerns mean that they have to be closely related to the mathematics national curriculum to earn their claim on the school's finances.

In various parts of the copse there were ropes, pulleys, tyre swings, diggers, buckets, jugs, bowls, logs, benches, a fireplace and trivet, piles of sticks and branches. Eight 5-year-old children arrived and sat on logs around the edges of 'the green rectangle' groundsheet. On the groundsheet were sticks about one metre long and wooden disks about 8 cm in diameter. The lesson is described in Table 8.1.

In teaching such as this, an input-output model of teaching and learning cannot apply. Instead lessons involve complex, contextualised experiences of multiple concepts, some are planned and some arise spontaneously. I would say children are learning to be mathematical, to think mathematically, in ways that can arise in the whole of life and are impossible to replicate with micro-managed lesson scripts. The use of language is complex, because it is a combination of the planned use of mathematically

Table 8.1 A Forest School session for year 1

Lesson episode	Mathematical content
Sandy drew an open-edged 3 by 3 grid for use with 'noughts and crosses'. In unison, with Sandy pointing, they counted the cells; some children observed one-to-one correspondence but others chanting numbers and not stopping when they got to 'nine'	Purposeful enumeration
She played one round of the game with one of the children to show how the aim was to 'make a line'; hers was a diagonal	Straight line using three points
She asked them to decide how many sticks to take to replicate her drawing. Some jumped to saying 'nine' (presumably because of the counting they had just done); others chose to replicate before saying how many. Sandy intervened when the replicated grids did not appear to have equal-sized cells. How would the sticks have to be moved to make the cells more equal? She referred to her own diagram saying 'like mine'	Enumeration Connecting edges and equal areas Relating diagrams to physical objects
Sandy then told them they could use the discs as the 'noughts' but would have to make their own crosses with little sticks. They needed to find thin sticks that 'measure the length of your hand'. Could they make enough to make five crosses? Some collected sticks that were 'too long' or 'too short'	Measure with non-standard units. Comparing shorter/longer; less than/more than
They played the game in pairs. I did not hear any discussions about strategy. In the games, I watched their emphasis was on getting your own line. Some children did not recognise a diagonal as a 'line' even though that was Sandy's example. Turn-taking was talked about	Straight lines were implicit

'Free play' was announced. Many activities then took place where mathematics was implicit and discussed when adults were present, such as balancing (how can the two of you get onto that tyre swing?), pouring (do you think you can get all this rainwater into that bucket?), sharing, measure (how high up the tree are you allowed to climb?). These were solved by experiment. Other issues arose: identification of toadstools; digestion habits of worms etc.	Estimations of lengths, heights, capacities, masses, and several scientific, social and linguistic concepts
Hot chocolate time took place with everyone sitting on benches in a rectangle. Sandy asked someone to count the number of children and all to count the number of stacked mugs. Again some merely recited numbers so she pointed to them one by one to emphasise counting. There were 12, so she asked 'How many will I not need to wash up?' After some discussion she set out 8 mugs and they checked to see that there were 4 left over	Enumeration Difference Conservation of number
Shall we have full mugs? Someone said 'medium' and Sandy restated this: 'half full'. The mugs were transparent so they could see how full they were. 'Tell me when they are half full – when to stop'. 'Is this less than/ more than a half?' 'Are they equal?'	Estimating halves. Less /more than Equal capacity in congruent mugs
After that ten minutes more play was announced, Sandy held up both hands to indicate ten	

useful language and responsiveness that comes from listening to children—listening for particular words and comparisons and listening to children's expression of their own observations. The emphasis was on their talk, rather than formalising mathematical ideas. With older children there may be some talk of strategy and more focus on anticipation and reasoning. This is in addition to social learning and the growth of confidence in being in a natural space with its noises, textures, dirt and dangers.

Sandy honed spatial and dynamic contexts from the land to generate mathematical experiences that related to natural ways of behaving. I was interested that the issue of connecting number words to cardinality using one-to-one correspondence arose. Counting and one-to-one correspondence find the quantity of discrete objects but measuring, comparing and estimating find quantities of continuous and many natural objects. This is similar to many of the teaching ideas in Indigenous cultures (see Chap. 6) and some home-school practices (see Chap. 7). Not all children had made the connection between reciting number words and stating a quantity but they could all compare quantities somehow.

Ed, who was home-schooling in the previous chapter, is also a Forest School practitioner, employed part-time by a primary school to give all ages the Forest School experience once or twice every term. The principles of his work are slightly different to Sandy's, in that his sessions are always led and shaped by the children's interests so are not connected so closely as hers to the curriculum. His main focus is on their social, emotional and physical development throughout their time at the school. However, having an engineering background himself he is aware of the mathematical principles behind much of their activity so that he can insert questions, direct attention and engage them in thinking and talking about measurement, spatial structures, capacity, angles, position, mass and forces where these relate to what they want to do or discover. He says that application of mathematical ideas occurs spontaneously, such as using the language of angles when asking whether people can see each other. I felt that calling what he does 'application' was probably underplaying the contribution to learning, since many mathematical concepts depend on concrete experiences and images to draw on when the time comes to formalise.

A Mainstream Primary Class: Gathering and Structuring Knowledge

So far in this chapter the settings have not been main classrooms, but specially structured groups that have a strong element of care for individual students as well as care for their mathematical learning and are funded to do this kind of work within the maintained system.

The year 5 class taught by Nicola is the first mainstream class in the book. She teaches 9- to 10-year-olds in a school that serves an area of a city that consists mainly of social housing. There are many children who come from households with low aspirations; many parents can be defensive or aggressive when dealing with conflict or with their children's problems. Home language is often restricted with children not having the opportunity to use language for thinking and reasoning except in school. This affects their cognitive development negatively so across the whole school there is a focus on developing language to support reasoning and problem-solving. Nicola has taught for 23 years and finds this community the most challenging in which she has worked. She has support from teaching assistants with whom she has worked for a while; mostly their approach matches hers. The focus of her mathematics teaching is to help the pupils learn how to learn, to gather and organise their knowledge so far, and to think more deeply about mathematical tasks. There are many weaknesses in their mathematical understanding and also in literacy. Her view of mathematics is of continuous strands of meaning that grow from early years to secondary school, so although the NC is written as a year-by-year document she starts new topics from earlier concepts and experiences. Their school experience may not have been continuous and their mathematical experiences may not have created a coherent pathway. When planning to teach she anticipates what might be familiar from earlier work and finds out what they know, which is sometimes more but usually less than they think they know. Her planning emerges from this to address the images, language and questions that are associated with a mathematical idea. In doing this, she uncovers what they remember and how it is structured. In her own teaching, she works explicitly on developing their memories by using rhythm, chanting, visual images and episodic prompts such as 'remember when …'. They tend to want to know what to do in response to tasks; they want obvious prompts to action rather than being prepared to think and reason about mathematics and decide what to do for themselves. Her care for their learning is informed by an understanding about how long-term memory for mathematical ideas is constructed from repeated experiences with multiple representations and access routes, and she has to contribute to students' formation of complex and useful concept images on which future ideas can be built. Access to these images has to be via fluency in language, facts and methods that trigger memory, so she has instituted her idea of a short daily 'Maths Meeting' (as advocated by Mathsmastery.org.uk) in which whole classes focus on repeating some of the core ideas on which future fluent learning depends.

In her teaching, which responds to what she finds in these initial stages, she is constantly checking for comprehension and if someone does not understand she will 'drill down and not let go' by spending time 'rewording, redrawing, reshaping, requestioning' until they do. This she does for one individual within the whole class context so that everyone can benefit from the extra chance to work at the idea and also can see that she cares about whether they individually understand or not. This is not always successful. For example, she recently taught them about square numbers, using diagrams of squares, finding the numbers on a multiplication grid, emphasising the language and symbolism and even going on to talk about cube numbers, but when this was followed up with assessment questions many confused them with multiples of two. The presence of '2' can trigger something more elementary, and a tendency to act quickly rather than consider mathematical meaning hampers response to contextual mathematical problems, so she slows them down and models good reasoning by focusing first on reading the question using techniques they use elsewhere in school, that is highlighting key words before tackling meaning. They then construct an appropriate number sentence, and she refers to visual structures for helping with this, bar models or numberlines depending on their past experience. In these ways, she introduces them to working more deeply on mathematics and enriches their memories with more coherent and interconnected images and language. She demonstrates that she cares about their learning implicitly in her flexibility and explicitly in her classroom interactions, including working with groups of flexible composition with various purposes, sometimes supporting, sometimes challenging. Her philosophy is that her teaching should address the way they learn, rather than the other way round. During her career she has realised that, whatever the current discourse for improving learning is, most depends on the teacher's belief that children can learn and on passion (her word) for their learning. That word crops up again, passion, used by Ruth Merttens (see Chap. 5) to describe the quality of attention a teacher can give with learners to mathematics.

A Mainstream Class: Students Choosing Their Difficulty Level

The teachers in the next few sections are working with students who are at the end of childhood, the beginning of adolescence—a time when they still may need support to develop personal power. They recognise that confidence, conscientiousness and appropriateness of choice are affected by emotions and social pressures and peers as well as by individual judgements of competence, so helping them learn to make decisions about when to be comfortable and when to be challenged is also part of their work.

In a year 6 class, students were given a high level of responsibility for monitoring their own learning and for maintaining the learning and emotional ethos of the classroom. For example, at any time a child who felt they had a concern that affected everyone could stop the whole class and speak to them. Circle time was used regularly for a range of purposes, and this could be initiated by the teacher, Jennifer, or by a child after consultation with a teacher. When taking their seats for circle time, when lining up for lunch, and when gathering on the carpet, the teacher asked them to check whether they were mixing their genders, racial backgrounds, friendship groups and so on. So how did these enacted values relate to their mathematics?

The mathematics lesson began with the teacher announcing that they were going to work towards subtraction with three-digit numbers, and this would be a 'peel-off' lesson. To start with, all students were seated on the carpet around Jennifer who wrote on a flipchart. She reminded the class about what they needed to do. They needed to watch and listen and ask questions if they did not understand. If she got to a place where they knew they needed to practise, they should 'peel off' from the main group and go to their tables to continue working on that stage of the work. Materials for practising related tasks were arranged round the room as were various concrete representations of number counters, linking cubes, Cuisenaire rods, base ten apparatus and so on. If they felt confident they stayed with the main group until she finished her presentation including harder examples. The teaching began with some quick practice in number bonds because those might be needed. Then there was an example of using single-digit subtraction in a column layout. Then came some two-digit subtractions in columns without and with decomposition. At this point, some students started to 'peel off' to work on their own, or in pairs, or went to a teaching assistant for help. Then there were examples with three digits some of which needed decomposition. A few more students left at this point. Further examples concluded decomposing across more than one column and, later, some examples with zero as the middle digit.

The follow-up work with materials allowed for several kinds of engagement with the meaning of subtraction, even where students were not fluent with number bonds. The teacher input had been necessary for illustrating a formal written layout. For example, while there are other ways to perform the subtractions with materials, the column decomposition method requires fluency of number bonds to make it worthwhile, otherwise it becomes tedious and overloaded with steps to remember. In an atmosphere of responsibility for personal effort and classroom ethos it seemed to me that there was a mixture of directed attention and autonomous sense-making. These young students could take responsibility for their own progress, and also for the social and emotional life of the classroom, in ways that gave them some control over their mathematical experiences.

Jasmina, a secondary mathematics teacher working in an urban comprehensive school, developed a similar 'choice' mechanism when providing her class with practice in algebraic manipulation. To give students personal responsibility for deciding the level of difficulty they needed she split presentation slides vertically to contain progressively more complex examples from left to right. The structure of the sequence of tasks was:

Slide 1: one simple example (blue)
Slide 2: worked answer to simple example
Slide 3: split into two parts: on the left further simple examples (blue), on the right more complex examples (green)
Slide 4: answers to slide 3
Slide 5: split into three parts: blue and green examples as before, yellow examples with extra complexity.
Slide 6: answers to slide 5. This is followed with a fourth set of examples.

With this method, students have rapid feedback in the form of worked answers and can decide when to move to the right and do more complex work. This is her pragmatic solution to her inability to monitor and give individual feedback in classes of 30, while also giving some choice and responsibility. She helps them maintain motivation by her presence, her encouragement, her one-to-one moments in lessons. She listens for what they say about the mathematics, how they describe it and how they see it, that will differ between students. Some find it hard, she says, to 'see' abstract ideas through diagrams as they are used to a video world in which perception is immediate and does not require interpretation. 'Reading' visual stimuli with meaning can be a difficult step. Her role, in part, is to adapt her explanations to these different capabilities.

In both of these classrooms, care for students involves them making independent choices about using procedures modelled by teachers. In Nicola's classroom, the teacher takes responsibility for diagnosing their current state of knowledge and designs her teaching accordingly, using multiple representations, materials and talk to develop understanding and memory. In Jennifer and Jasmina's teaching the student takes responsibility for choosing what to do, but in a context of being sure that teaching will respond and adapt to their ways of understanding mathematics. These are different kinds of care for learning, arising from different perspectives of mathematics: procedures to be carried out with the help of materials and diagrams, or concepts that can be experienced and understood through materials and diagrams.

A Cluster of Primary Schools: Care for Mathematics as an Institutional Concern

The teachers so far are working either in specially funded circumstances, or in ways that are not necessarily common in their schools.

I found a small group of primary schools in which there is a shared philosophy of care for students and care for their learning of mathematics throughout, supported by multiple forms of in-house professional development and a strong inclusive commitment and leading to above -average national test results. Several senior staff have post-initial-training qualifications in mathematics teaching and work together to develop ways to promote inclusion and mastery in their schools. The meaning of 'mastery' for them is about flexibility based on dialogue within whole classes with the guiding principle that every person has the right to high-quality mathematics education in the classroom among peers, and the job of the teacher is to give all students experiences that realise this fundamental right. The schools serve populations that contain a high proportion of students with a range of psychological, social and emotional needs as well as cognitive delays, this can be as many as 50% of a teaching group. The work was reported to me by Sarah, who commented that: 'If we can get our teaching right for the most vulnerable, it is going to be right for everyone'. Faced with 30 brilliant students (her words), who have all been learning since they were born, how do teachers control what they do themselves to ensure these 30 people learn mathematics? The responsibility lies with the teachers who are working within a structure that makes a 'universal offer' available for everyone, 'targeted offers' which are additional accelerating interventions three times a week for those who can access the universal offer with some extra input, and 'specialist offers' for students with complex needs requiring personal, expert, support. However, for the main mathematics lessons they are all together in a shared conversation between the teacher and the whole class.

How does this work without the teaching somehow pointing to a notional middle group of students? Sarah supports the development and use of a common pedagogy throughout the school. She describes some lessons as being about 'small steps' that are not bits of procedures but essential conceptual components that connect with other components—connections being a key focus of learning. In such a lesson the teacher presents an idea, and there would then be some pairwork and individual thinking followed by students giving feedback to the teacher about what they have done, thought or understood. The teacher's job is to listen to what is said and take it into account when deciding what comes next, whether that is a one-to-one discussion with that student from which everyone might benefit, or a slightly different task, or an example with some different features, or a reminder of something met earlier. Talk is very important, both for informing the teacher about what they understand, or about the teacher modelling speech patterns or directing students. For example, students are encouraged to say: 'I have noticed …', 'I found out that ….', and 'because ….'. There are no long periods of working on several similar examples; all examples and questions have a conceptual purpose and if anyone in the class is having difficulty then everyone needs to discuss why that might be. A teacher might respond to an unexpected comment by saying 'That's interesting. Let's discuss it' or 'that's interesting, we'll come back to that next lesson'. The aim is to develop ways of learning mathematics, and to thinking critically about situations, as much as it is about accumulating knowledge and procedural experience. The teacher strips a lesson down to essentials to concentrate on their meaning-making with these essentials and identify critical features that need more attention before trying to build further on the key idea. In this way, they all feel valued and heard and understand that she will use their perspective to shape the lesson. Sarah talked about how timid children become empowered by this process and are able to 'own the space' and become generally more confident and successful.

There is another kind of lesson as well in which mathematics can be learnt in collaborative, exploratory and applicable situations. These are interdisciplinary team tasks in which realistic problems can be solved and artefacts created. One example is to design and make a waterproof shelter. As well as involving the use of generic problem-solving skills such as asking: 'what do I know?' this kind of task creates a need for mathematical knowledge and tools, and these are often the first experience of a new-to-them idea. For example, a project on flight led to building scale models of various flight machines in year 2, where ratio and scaling do not appear in the curriculum for a few more years. Similarly, the shelter task led to discussions about how to quantify three-dimensional capacity. This approach means that later mathematics lessons that work more formally on such ideas can start with 'do you remember when…?' and draw on episodic memory, physical representations and the scope and limitations of a concept. The mathematical journey of each individual student is supported by referring back and sideways to successful experiences while moving forward into extensions and expansions of ideas.

Sarah mentioned 'measure' and quantity as particular ideas that permeate mathematics. I have referred to this many times in this book since experience with materials enables students to make sense of mathematics, rather than the abstractions that take place when calculating or manipulating symbols.

Secondary School Lessons

Is the kind of care for understanding shown by Nicola possible in secondary classrooms, where the pressure to get students through qualifications heats up, adolescence kicks in, and teachers might encounter 120 or more students during the course of a day?

Morgana

Morgana is an early career teacher in a large secondary comprehensive school in a Northern city. It serves a population with a social and ethnic mixture of backgrounds, and mathematics results in the school are above the national average in a school context where overall results are not so successful. She came into mathematics teaching after achieving an A-level herself, followed by a degree that had a little implicit mathematical content about structure and spatial reasoning. In adult life, and as a mother, she kept her mathematical interests alive by studying at home and working on mathematics alongside her own children as they grew older, eventually following a subject enhancement course before qualifying as a teacher. Her care for students and learning is evidenced in multiple ways.

She is available to students at all times, and her classroom is known as a safe and welcoming place at lunchtimes for 'hanging out' and also for one-to-one talks to support mathematics and also for emotional support. Her relationships with students are warm and encouraging, both for their everyday problems with peers or at home and also for their efforts in mathematics. She described one student in year 11 who has recently been placed in care. This is having a profound negative effect on his GCSE expectations—she provides him with tea and biscuits, a safe place and encouragement. She says she 'never says no' to students asking for any kind of help.

Although students are taught in sets, she tells them not to accept the limitations that being in a certain set might convey, but to aim as high as they can. To seed this 'aim higher' approach her lessons include tasks that push students out of a comfort zone to do harder tasks. She supports this with questions that frame their efforts and at the time of our interview she was actively extending her repertoire of such questions and gave me her current list:

What does this question remind you of?
What other topics can you link it to?
What happens if I do this to the question? Or add this into it? Does this help in anyway? If so, how and if not, why not?
Why don't you jot down what you do know from this question, what parts you can answer and then build around that?
Why not ask the person sitting next to you how they would/have approached the question?
Why don't you highlight key words - look at the topic surrounding this question and write down what you know. Now use this and apply it to the question in front of you.

Her aim is for them to do the thinking involved in linking mathematical ideas together so that they have a strong networked basis for future work. The tasks themselves come from a whole-school resource bank, a scheme of materials, and from colleagues. The department has two meetings each week during which approaches to topics are discussed and Morgana feels she is an equal participant in a team that is collaborative and shares similar aims. The department allows her the autonomy to decide when and how to teach and the flexibility to move through the curriculum at a responsive pace. Much of the class time involves discussion with the whole class about sense-making, students listening to each other, her listening to them and deciding how to modify and develop her plans accordingly. She explains to them why she asks them to work in certain ways, such as when to think silently about some mathematics. In some ways, her lessons can be uncomfortable if students expect her to give answers because she always expects answers to come from students, after some effort, and not from her. This expectation is mediated by their own very vocal comments if they do not understand; in general, her students are not quietly compliant!

Each new topic starts with an audit of their previous understanding: how and in what ways concepts have become embedded in earlier learning. Even though a long time might be spent on one aspect, such as place value in year 7, lessons are not boring and repetitive because different manipulatives, representations, and ways of reasoning are used to build strong understanding on which future learning can build. She comments that this kind of teaching has led her year 7s and 8s to be more confident and competent with angle reasoning than her year 11s whose original learning was fast-paced and superficial. She says: 'I want them to love the subject and go where their curiosity takes them. It enhances my own relationship with mathematics'.

She is not in line with the school behaviour regime using a system of warnings that lead to automatic detentions and an escalation of consequences for poor behaviour. She therefore makes her own decisions about whether, when and how to send disruptive students into that system and claims she is known to be 'soft'. Her own behaviour system is based on the expectation that adolescents can reflect on their behaviour, analyse it, repair damage and redeem the situation if given the opportunity to do so. Only after further problems will she trigger the detention system. However, she has no faith in the detention system as doing any good, citing some students having 'block bookings' into the system after a day of disruption, and invites students for lunchtime discussions, which might be about their understanding of the lesson content, or other reasons for their behaviour. It is better to talk, she says, and build mutual trust by giving each other respect, restoring a human element to teacher–student interaction. There is tacit agreement from senior colleagues that her approach 'works'. Although she claims to be soft, she does draw red lines around behaviour that directly harms other students, such as the use of hate language.

Tom

I observed Tom, who teaches in a comprehensive secondary school, teach two lessons to years 7 and 10. It was at the beginning of the Covid-19 epidemic and the country was rapidly going into a period of heavy restrictions on movement and behaviour; school routines were disrupted and attendance was falling. The year 7 class was only two-thirds of its normal size. There was some student chatter about who was absent, and who needed to move to find someone else to work with, and the sharing of virus gossip. The year 10 group settled more quickly to work, with nearly full attendance and a studious atmosphere. These differences in behaviour could be explained by what is known about adolescent foci at different ages, older students being more future-focused and hence purposeful, while the younger ones are more peer-focused. Tom's behaviour towards students was similar in both groups: friendly, quiet, listening, responding, using names, and showing an interest in their mathematical thinking. His students knew they were valued through respectful dialogue.

Tom told me that his planning involves imagining a sequence of lessons and tasks as the unfolding of a narrative that has character, plot, motivation and resolution. This is not a linear sequence of events along a timeline with the intended mathematical idea as the end point but a story for which various viewpoints, assumptions, bridges, wonderings, justifications provide the paragraphs and the plot might be a braid of components, sometimes with parallel strands. While planning he is asking himself: 'Where are we trying to get to? What are our resources/tools/knowledge? What might we do? How can this be pulled together?'

In the year 7 lesson Tom was working in the medium term towards the students being able to use algebraic representations of operations, and to solve linear equations. In the longer term, he is supporting their reasoning from self-generated examples. All students need to develop the intellectual resource of argumentation, and this fits with an increased desire in early adolescence towards personal arguing power. The first task was to sort a collection of statements he had designed about number operations into those which are always, sometimes or never true. Here are a few of them:

Adding 6 to a number and then subtracting 5 from the result is the same as subtracting 11 from the original number.
Adding 78 to a number and then adding 57 to the result is the same as adding 135 to the original number.
Adding a number to itself is the same as multiplying a number by itself.
Multiplying a number by 5, then dividing by 20 is the same as multiplying by 4.
Multiplying a number by 6, then adding 5 is the same as adding 5 then multiplying by 6.

This kind of task helps students to delineate the boundaries of truth domains, and also introduces a need for proof. It also allowed him to introduce the idea of quantifiers 'for all' and 'for some' and 'impossible' informally.

Students had previous experience of wanting to rush to conclusions based on one or two examples and they had discussed this as a class. While they worked on the statements Tom worked his way round the room talking to those who seemed stuck, or seemed confident, or who made interesting remarks in his hearing. His main way of approaching was to squat down to table level and ask questions such as: 'What did you think at first?' 'What did you think about …?'; 'What do you mean by …?'; 'So what are you saying about it?' Each of these questions gives access to their thinking and/or directs attention to aspects of their thinking so far. He then asked the whole class to offer their thinking about statements they were sure about. The first was the middle statement, and a student offered 2 as an example where this statement is true. Someone else then offered 3 but it wasn't clear whether they understood what this example contributed to the overarching question of whether it is always, sometimes or never true, or whether it had been offered merely as another example of doing those operations. The next statement chosen was the fourth one and the student who volunteered this said '5 times 4 equals 20', giving as a reason that it looked right. Another student then offered $20 \div 5 = 4$ as justification. After some discussion Tom asked: 'I start with a number, I multiply it by 5, then divide by 20. Is that the same as taking the original number and multiplying it by 4? If you are saying this, then you are saying that the statement is always true.' One student said 'maybe sometimes' and he asked them to vote whether or not they agreed. Nearly half agreed. It turned out that some were happy to agree that $5 \div 20$ is the same as $20 \div 5$. Tom's afterthought about this whole incident was that they were naïve about what an example could tell them, and that this may have arisen because all their previous work with student-generated examples had been to create examples to work on, and this may have been the first time he had expected them to think about the behaviour of examples and the relationships within them. Overlaid onto that might be a voting habit that is about preference-in-the-moment rather than reasoning (see a different voting habit in Saul's practice below). Furthermore, their willingness to agree could have been built on an assumption that commutativity is the norm for binary operations, plus a recognition that 4, 5 and 20 are related in a multiplicative way—hence it looked right. Each of these tendencies had overridden the main task about truth, proof and disproof. Making a mental note to do more work on multiplicative relationships he moved on to another task, still related to operations, and this was more successful—they had to find and explain equivalent expressions in this collection:

$$
\begin{array}{cc}
5 \times 9 & 9 \times 30 + 5 \\
(10-1) \times 35 & (40-5) \times 9 \\
9 \times 40 - 9 \times 5 & 9 \times (30+5) \\
(10-1) \times 30 + 5 & 9 \times 30 - 9 \times 5
\end{array}
$$

In the year 10 lesson Tom went beyond the curriculum into an area of more advanced thinking. The topic was sequences, for which the curriculum expects linear, quadratic, arithmetic and geometric and some others; he used rational functions because his earlier teaching had included the notion of limits of sequences and he wanted to continue that aspect of the story. Students had to construct the first five terms of each sequence and discuss whether they converged or not, and if so to what. They then looked at $\frac{3}{(n-2)}$ and discussed the effects of the 3 and the −2 on the limit of convergence. While Tom orchestrated discussion he did not evaluate what students said, he let them continue until students stated that '3 divided by million is the same as 3 divided by a million minus 2' and 'it cannot go past zero'. They then sketched graphs using their first five terms and knowledge of convergence and, as a whole class, responded to questions such as 'what happens to y as x gets very close to …?'

They then looked at graphs of several, mainly rational, functions and he introduced them to a notation for the asymptotes, focusing not on infinity, which is the usual way in which teachers excite students about asymptotes, but by distinguishing between whether a limit was approached, by which variable and from which direction. One of the functions to which they would return in the next lesson was $\frac{3}{(x^2-9)}$ and Tom was planning to ask them to think about values around $x = 3$.

In Barbara Jaworski's terms (see Chap. 5), mathematical challenge beyond the immediate questions was identifiable in each lesson with support for enabling students to engage with the challenges. In both lessons this support was verbal, example-based, and developed through dialogic argument involving the whole class. The groundwork for developing a thoughtful ethos was done in smaller interactions as he went round the room during pair work. He asked questions like 'what did you think when you read this?' and 'what do you mean by …?' with both classes. Like Morgana above he is always asking questions rather than giving answers.

I was particularly struck by the sense that, although he had designed the tasks and had an overall sense of the plot, the narrative itself (his word) was the responsibility of students, whose thinking was his constant focus. It is their own narratives that constitute their learning. In the year 10 lesson, although he was the mathematical expert, there was a strong sense of being alongside students as they constructed their narratives.

Octavia

Octavia teaches in a suburban comprehensive that has a level of achievement in mathematics comfortably above the national average. As well as being a qualified mathematics teacher she has learnt *Philosophy for Children* and *Godly Play* pedagogies, both of which focus on what learners think about abstract ideas rather than their acquisition of knowledge, and has taught in primary schools and also in special educational needs provision. Here is a description of a lesson that illustrates several deliberate strategies that she has developed as typical of her teaching.

The year 9 lesson started with an instruction to draw a 3 × 4 rectangle on squared grid paper and to draw the internal lines to make a 3 × 4 square grid. She starts most lessons with something like this, an opening prompt, diagram, calculation or image, and then asks them for five things they notice. This means that everyone could get started with some thinking about it. The question for this image was: how many squares are now outlined within the rectangle? Typically, some students stop at '12' which is quite a simple idea for them, while others who have adopted the practice of going beyond what is obvious also identify 2 × 2 and 3 × 3 squares. When these ideas were shared there were new ways available for everyone, including Octavia, to look at the rectangle. Octavia then directed them to look at how many straight-line segments are needed for the grid. The class then searched for general relationships between the number of straight-line segments and the number of squares in rectangular grids. This could be done in a variety of ways. For example, one group of students restricted themselves to squared grids and organised their counting in such a way that they developed a conjecture that it was not possible to construct a grid that would give them a total of 100 squares.

Octavia's aim for this lesson was for them to use and develop their inquiry skills productively. Different groups used tables, graphs, diagrams, knowledge of square numbers and their experience of constructing the *n*th term of sequences in different ways, thus coordinating their own arguments in different orders and with different priorities. Whatever their individual pathways, for all of them this became an early experience of partial sums of sequences and of the development of conjectures and proofs. Classes that have been with her for some time understand that effortful thinking is expected of them. Classes that are less acclimatised to her teaching sometimes find it hard to adapt, particularly as there is nowhere to hide if they would prefer not to engage with the mathematics. Rather than browbeat them into her ways she has sometimes had to adapt her teaching so that it is closer to what they have had in the past (being told more; imitating and practising some methods) and gradually shift towards them taking more responsibility for engaging with tasks and hence becoming able to use the freedom that she offers them purposefully and powerfully. In her school, she is the only teacher to work like this but her students' results in national tests give justification for this approach. She is educating them to think about mathematical ideas.

This way of teaching reminded me of classrooms that used investigative approaches to mathematics that blossomed in the 70s and 80s. However, Octavia is too young to have experienced these and it seems that she has developed her teaching philosophy independently, drawing on her experience of the cognitive capabilities of younger children and those with learning difficulties as well as her own high level of mathematical knowledge. She is also aware of the intellectual potential of adolescents and their need to relate to a peer group, as well as longer term needs to do well in mathematics. She has sought and discovered sources of tasks, such as the rectangle grid task, and sources of professional dialogue that support her to work in this way.

She was very clear about her understanding of the nature of school mathematics—that it needs to shift away from a sequential procedural approach towards an object focused approach. She refers to computer science as having made a parallel shift in recent years. Students, in her view, need to find out what can be done with concepts, to experience the scope of procedures, and they do this best if they are given abstract ideas through concrete and symbolic situations that are manipulable and can be played with safely. Her role as a teacher is to locate the essence of mathematical ideas and present them in accessible tasks. Her next role is to listen to students so she hears what they are aware of and how they decide to process the mathematical objects. Even in more straightforward retrieval and practice tasks she notices that they make connections she could not anticipate, so her understanding of what is possible in students' minds grows. She has recently begun to use more 'true/false' questions because these give her information about how they see the scope of a definition or procedure. However, while precision and mathematical integrity are important she also accepts statements and definitions that are true in a limited domain if they are useful for the current students in the current mathematical context. The awareness of different ways to see mathematical ideas means that it is always possible for a lesson to go in an unexpected direction, even when she is working to a detailed personal script and her students are learning and practising procedures. Typically she will observe what they are doing and select students to write their solutions on a public whiteboard, whether right or wrong, and then ask the whole class to think about a particular solution, share their thoughts with a neighbour, and then with the whole class. They are used to this collaborative way of extending their repertoire of methods and learning to avoid pitfalls. She uses the 'think-pair-share' strategy often as a device to engage all in the collaborative thinking and to build confidence.

Through listening she also becomes more aware of their barriers to learning. She is very careful in her use of language, noticing that there are always some students who interpret tasks differently—even small changes in language can make a difference to how students perceive a task and make decisions about it.

What is remarkable is that she has developed these methods not by copying the pedagogy of others but by thinking about how to apply her other teaching experiences to her classes of 30 while improving the learning of all students. She has what Schoenfeld (2014) would call 'a powerful classroom', that is one in which the content is true to mathematical discipline, where tasks make cognitive demands on students, where everyone can engage and think of themselves as mathematical people, and where their own thinking is out in the open (she refers to TRU (https://truframework.org) as an authority). Authority and direction can shift to students rather than remain with the teacher. Her choice of tasks requires students to think effortfully; their thinking does not depend only on her questioning during or after the task. However, the direction of content focus stays with Octavia.

Saul

Saul teaches in a secondary school in a small town in an area where there are two grammar schools, so many students who come to this school will have applied but not been selected for grammar schools, or already thought they were not clever enough. However, the curriculum expectations and the success criteria by which schools are measured are the same for all schools. There is a 'value-added' measure which is the progress students make from a baseline when they enter the school. Saul's teaching has significantly higher value-added scores than his colleagues so it is safe to assume that he is doing something exceptional within his school context. He gets mixed messages from senior management who are pleased with his 'value-added' scores and also talk of spreading his pedagogical strategies further across the school. However, he does not feel supported when parents complain about his methods. Thus, Saul is working within a cultural context, both within and outside the school, that expects more uniformity of pedagogy, and has an input-output-assessment expectation of education to which he does not conform.

A year 7 lesson of 30 students, who had been in the school for 3 months, was the second lesson they have had on square numbers. They would all have met these in primary school—possibly learning a list and something about them. He asked them to write down as many as they can. He then wrote them up to 400 on the board and asked them to check their own work. He then had an 'eyes closed' vote. 'Put your hand up if you did more than you did yesterday'. This is a general tactic he uses again and again to get a sense of their progress without them seeing other students' votes. He then asked: 'Who can write down the next one without doing a long multiplication? Look at the numbers again to see if there is a pattern and, if you can see one, use it.' They were likely to have experienced sequence pattern spotting in primary school but maybe only with linear patterns. They were silent for a while. He then told them a story about how he came to like sushi by trying again and again over time and finally 'getting it'. He gave them the responsibility to make the effort to learn and said he would give them 90 seconds more for silent thinking. After that there was an eyes-closed vote when he asked, 'Who still does not see a pattern?' He said: 'I still don't know how many of you have not seen a pattern so we'll look at it a different way'. He threw a pile of wooden cubes onto a table and asked them to crowd round so they could see. Some were reluctant to move to where they could see, presumably because of the friendships, hostilities and embarrassments of adolescence. He made it clear that it is their responsibility to make sure they could see so they could learn, asking them what they should do if they could not see. When he was satisfied that they could all see he slowly constructed, with exaggerated sequential movements, a spatial pattern with the blocks. In Fig. 8.2, I have numbered each component in the order in which he introduced them.

Every now and then he hid part of the shape, or the whole shape, and asked, 'What have I got behind my hand?' Eventually, after doing the same to construct a five-by-five square from a four-by-four square, he asked if anyone could come and do the same and emphasised that they should do *exactly* what he did: that is, add the two rectangular pieces and

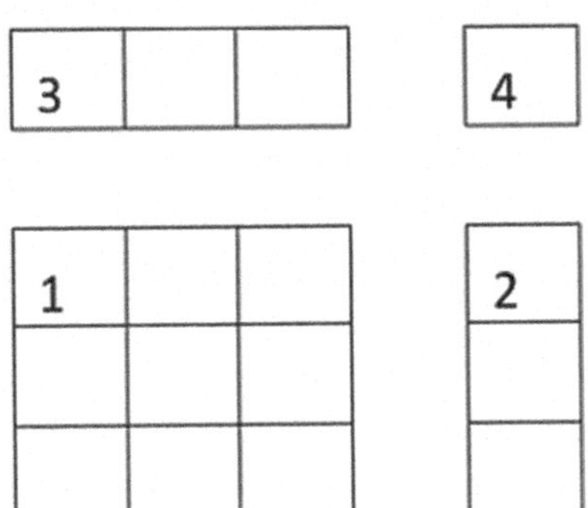

Fig. 8.2 Saul's arrangement of blocks to illustrate construction of square numbers

the final single square each time. There was some discussion and laughter about whether he expected them to do exaggerated movements like he did or not.

'Talk to the person next to you about whether T (a student) did it exactly the same way as I did it'. Eyes closed: 'Put your hand up if you need more time to talk about it … what's the pattern?'

During this sequence he did not say what the spatial pattern was that he wanted them to see—the responsibility was given to them; he had structured the physical and visual aspects in a dramatic episode that enacted $n^2 + n + n + 1$. He wanted them to make a link between the blocks and the list of square numbers on the board.

One student described a pattern he had noticed in the 'ones' digits; another had noticed that the differences are 3, 5, 7 … and Saul said, 'Think about how this pattern might be related to the blocks'.

Somewhere in the middle of every one-hour lesson he generally gives students a break of a few minutes to chat and walk about; often he takes them all out to walk briskly around outside. His argument is that this a physical break is necessary for them to be able to think hard for the rest of the lesson. I saw three lessons in which this happened; it was a very cold day but the practice was accepted. After the break another student said: 'You have to add 2 to the previous number you have added', and Saul asked if anyone could say this without using the word 'add' twice. One said: 'The next odd number'. He then reminded them to think about the blocks and to imagine the physical movements. He drew a 20 by 20 square (labelled as such without any internal lines) and asked what he would have to add to it to build the 21 by 21 square. They had to think in silence. Then eyes-closed: 'hands up if you can see the connection' and the 'hands up if you only got it in the last 60 seconds'. From this he knew that one student has still not got it. He asked her to stay with him at the end of the lesson.

In a year 10 lesson he showed sustained insistence that the students should do the thinking. They had been working on angles and bearings, during which time he had discerned several aspects of mathematics that needed more attention, one of which was the solution of equations and another the visualisation of right-angled triangles in a bearings diagram. He had also noticed that most of them had a fear of fractions that meant that working with trigonometric ratios was problematic.

He decided to return to solving some equations that they had worked on recently so they could become more confident when fractions appeared and wrote these on the board:

$$x = \frac{8}{15} \qquad 2x = \frac{8}{15} \qquad x^2 = \frac{8}{15}$$

$$x+3 = \frac{8}{15} \qquad 3x = \frac{8}{15} \qquad \sqrt{x} = \frac{8}{15}$$

$$x-2 = \frac{8}{15} \qquad 3x+1 = \frac{8}{15} \qquad \sin x = \frac{8}{15}$$

He said he was interested in how they worked them out, not in the answer. There then followed a discussion about the difference between 'isolating *x*' and 'finding a value for *x*'. He kept asking them for the difference and they were expected to listen to what each other said about this. He said: "They are close but not the same thing'. Eyes closed: 'How many of you think you know the difference?' Not many put their hands up so he gave then more silent thinking time and then asked them to talk to each other about it, while he listened to some of their conversations. He found that their fear of fractions was still getting in the way, so he wrote $x + 2 = 10$ and said he did not want to know the answer, he wanted to know how to isolate x. One student gave a thorough explanation about getting *x* to be on its own on one side of the equation by subtracting 2 from each side. Saul explained that he had given them fractions because he wanted them to see that the presence of fractions did not alter the method. He changed all of the fractions to 5, except for the last one which was left as $\frac{8}{15}$, and the class as a whole described how to tackle each of the equations by using inverses. He then asked how they would isolate *x* in the last example, and several said 'inverse sine'. To me, it looked as if they were on an edge between thinking of solving as a type of calculation, hence they were happier with integers, and solving as restructuring that could be done with any number or expression. When doing trigonometry they had to be comfortable with rational numbers appearing in equations.

They then had a mid-lesson break and when they came back there was a bearings diagram on the board with a questions of the type 'given the bearing of B from A and C from B, what is the bearing of C from A?' and 'given distances between A and B, and B and C, what is the distance from A to C?'. I shall skip some details of the discussion during which I noticed that these students seem comfortable spending a long time silently thinking and he seldom uses 'hands up' to get them to answer. The problem he had discerned was with the need to construct right-angled triangles so that they could use Pythagoras' theorem, so he had cut out pairs of right-angled triangles for them to take home for homework with the task being to place one of them in the positive quadrant of a coordinate grid with one of the sides parallel to the x-axis. The second triangle had to be placed to that it shared a vertex with the first, and also had a side parallel to the x-axis. They were then to find the coordinates of other vertices of the second triangle. It is hard to see a direct connection, but he was hoping they would see the value of constructing right-angled triangles to find unknown measurements.

This lesson is interesting from several perspectives. He did not break the situation into simple steps for them, instead he provided insight into the tools they might use; they are going to spend several lessons working on the diagram because it draws together several concepts and the use of several procedural tools. He used this curriculum context to uncover, and work on, several underlying aspects of mathematical knowledge and reasoning.

I have described the mathematical details of the lesson, and some details of his pedagogy, because they illustrate his principles and also the tactics he has developed to enact those principles. I saw these principles at work in all the lessons:

- All can and should think effortfully; he gives them time to do so;
- All can and should learn; he helps them understand how and when they learn;
- They need to take the responsibility to be attentive;
- They can monitor how their understanding shifts; he gives reflection time for this.

- They are honest in the 'eyes closed' situations; he takes what happens into account;
- He provides multiple opportunities to make links between representations.

Thinking hard is a requirement in his lessons; he does not tell them how to think, or what to think, but directs them to what to think about. His teaching is intentional; they are not exploring without goals but if there are difficulties he takes his lead from them and works on whatever is causing the difficulty. The tasks I saw were planned to focus very precisely on specific mathematical ideas that had emerged in earlier lessons. In that sense, teaching is directed but also recognises that mathematical conceptual learning involves individual inquiry.

His responsive planning and teaching means that he is unlikely to 'cover' all the required curriculum in his lessons but he feels it is more important to develop their confidence in thinking about unfamiliar numerical and spatial situations rather than rush them superficially through procedural learning. Those parents who complain see slow coverage of content; he sees persistent growth of reasoning and substructure. I asked one student why they were not remembering what they had done in primary school about area of combined shapes and he said: 'Here it is much deeper'.

One further feature I noticed is his use of silence. It is not only used for independent thinking, but also as a short punctuation in the lesson when he thinks the class is getting fraught, or too many ideas are buzzing around. This is in contrast to Janet's practice of filling the lesson with words so that students with traumatic lives do not have to face inner demons.

Reflections

The cases in this chapter are all examples of teaching in which a learner's individual mathematical cognition is treated as an integral part of their whole selves and their current lives. The teachers provide an environment in which responsibility for, and control of, the direction of mathematical ideas flows between teachers and students depending on the

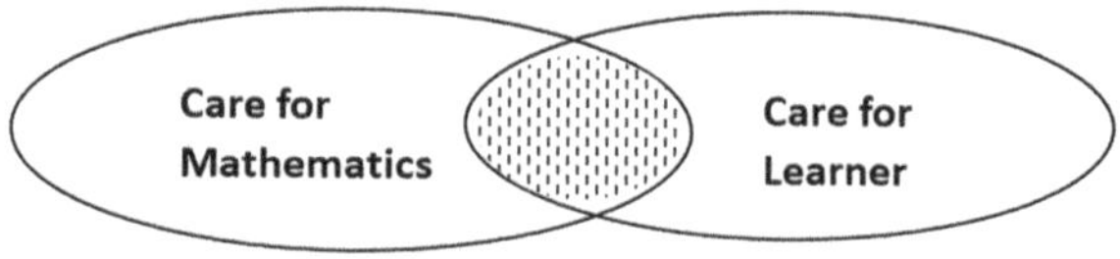

Fig. 8.3 Integration of care for learner and care for mathematics

mathematical thinking that takes place and the ideas that arise. The teacher prepares initial tasks and goals, adapting and extending as the lessons unfold. Interactions, both planned and spontaneous, support individual and class thinking.

Care for the learning of mathematics is manifested in many ways to populate the shadowy area in Fig. 8.3. In the final chapter, I will summarise features of practice that exhibit care for the learning of mathematics.

References

DfE. (2011). *Every Child Counts: independent evaluation*. https://assets.publishing.service.gov.uk/government/uploads/system/uploads/attachment_data/file/182404/DFE-RR091A.pdf. Accessed 11 Apr 2020.

Schoenfeld, A. (2014). What makes for powerful classrooms, and how can we support teachers in creating them? A story of research and practice, productively intertwined. *Educational Researcher, 43*(8), 404–412.

9

Care for the Learning of Mathematics

Integrated Care in Alternative Contexts

In the early chapters of this book, I explained that my interpretation of 'alternative mathematics education' is teaching perceived by teachers to be different from what they feel is taking place in publicly funded mainstream educational spaces around them. As I explored this meaning for 'alternative' it became clear that the teachers I found had developed aims and principles that permeated their work in ways that integrated care for students and for mathematics. Care was not only evident in their social and emotional work but was also expressed in the mathematical detail of their planning and interactions. I repeat here that I am in no way implying that other teachers, or teachers who teach in other ways, do not care. What has emerged is that it *is* possible to care for students and also to care for mathematics with integrity and loyalty to the subject, and to combine these through caring for the learning of mathematics.

A. Watson, *Care in Mathematics Education*, Palgrave Studies in Alternative Education,
https://doi.org/10.1007/978-3-030-64114-6_9

Funding and Class Size

Many of the contexts described in the book require special funding and staffing. In Chap. 8, the decisions of primary schools to provide Forest School, small group interventions and internal professional development for mathematics while budgets are cut to a minimum are inspiring. The argument Emily[1] gave me, based on her experience, that 'one good intervention is more effective than several smaller ones' is convincing. The provision for qualified nurture group staffing in secondary schools, rather than using stringent behaviour policies to tackle non-conformity, is also an inspiring alternative to exclusion.

Chapter 8 also contains examples of class sizes and staffing that are normal in English mainstream schools and that do not require special funding, yet have significant similarities with the underlying principles that emerged in Chaps. 6 and 7. Individual teachers enact care for individual students of mathematics in different ways: for example, not everyone can follow Morgana's practice of using restorative justice instead of school punishment policies but talk about difficulties with students' behaviour was always redemptive as in Lochville, ITS and Janet's classroom. What they talked about, and showed in their teaching, was care for their students' mathematical thinking.

Who Cares for the Teachers?

One teacher I met while preparing for this book had been told that every lesson had to be based on plans that the head of department had prepared beforehand; another teacher told me that all lesson plans for the year had to be completed during the summer vacation; another had been hounded out of a new job after a few months because he would not fit with an imposed scheme that limited responsiveness to learners. This book is about other possibilities, but stories about preventions like this are not hard to find in other places.

---

[1] Pages referring to individual teachers and schools are listed in a separate index of Teachers and Elders.

Before I talk more about care for mathematics I need to say something about professional development that is continuous and embedded in the life of the school. In Sarah's school, a shared team atmosphere supports all teachers. Because normal practice in lessons is responsive, any post-lesson marking of written work takes very little time. This means teachers can spend more time designing their next lessons together, thinking through their response to unexpected difficulties and anticipating what might happen. These activities give time and space in order to develop both personal subject knowledge and pedagogic knowledge and skills. Teaching assistants are supported to learn more mathematics, perhaps getting GCSE if they have not already achieved it.

Mark talked to me about the exhaustion generated by close attention to students' thinking, He puts his time and energy into planning for learning and classroom interactions. He was careful to tell me also that he felt he did not do as much in non-contact time as some others. For him, his attention during the lesson is at the heart of learning and that is why he calls his classroom 'thoughtful'.

Several teachers in the book commented on whether they feel different or similar to other teachers or whether they feel they are part of a team. Morgana is aware of being a relatively new latecomer to the profession but is treated as a full participant in the teaching team, despite having a different attitude towards behaviour management. Some of the teachers do not feel that they are understood as part of shared endeavour, but have autonomy because of their results. Others have considerable autonomy and feel respected by colleagues even though they 'teach differently'; some are leaders in their schools or have colleagues with similar pedagogy.

In ITS, teachers told me that a major difference in the work ethos from other schools was the expectation that teachers would be emotionally honest about themselves with each other. Without that honesty, how could anyone work together on the emotional health of children? Janet said something similar to me, but in her case she is honest with her students about how she feels so that they can learn more about how emotions work in groups of people.

Implicit in much that was said and seen was an impression that teaching is a reflection of themselves, rather than a performance. Lisa showed this in her determination to learn *from* the language of Mi'kmaq. Other examples include: teaching like Cap by working together with few explanatory words; Jennifer's use of circle time; Saul valuing silence; Jasmina learning about students' outside lives; Octavia's multiple sources of knowledge about how young people can express their thoughts and feelings; Janet's sensitivity to unexpressed emotions. One teacher, herself autistic, talked about how she has deliberately had to learn that not everyone thinks in the same way, and is able to talk with her students about this when necessary. In her work in Indigenous contexts, Cynthia Nicol identifies seven principles: reverence, respect, reciprocity, responsibility, synergy, interrelatedness and wholism (Sterenberg et al. 2010) and I see elements of these principles in all the teachers I have written about.

To my knowledge, they have all taken steps to maintain their personal mathematical learning in various ways since becoming teachers. Some have taken courses that increase their knowledge; some engage with workshops and online opportunities to work on mathematics problems with others; some are fortunate enough to work with colleagues who do mathematics together from time to time. Their care for their own mathematics informs their pedagogy.

They also have in common their engagement with various professional networks, either professional associations, or courses, or projects, or social media, hence they all seek contexts in which to extend their professional and mathematical knowledge. That was central to the way I gathered them for the book. More deeply, they seem to be sustained by passion, compassion, ethics, care, commitment to social justice and love. My interpretation of their work is that they prioritise students' thinking because they can do no other; they have no alternative.

Building the Contract

The environment and interpersonal relationships of the classroom comprise the contract between teacher and students described in Chap. 5 (Brousseau 1997). Mostly students learn that if they do what teachers ask them to do, they will learn the right mathematical knowledge and pass their tests. This is also what society expects of formal education. Figure 9.1 shows how these parts of the contract are generally related, with the teacher providing the ingredients of the milieu for learning.

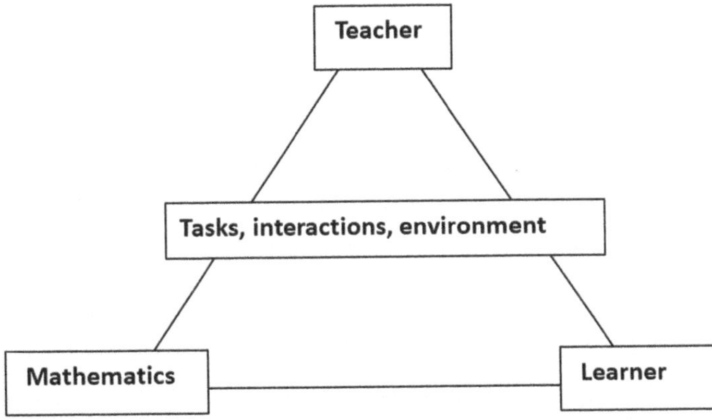

Fig. 9.1 A didactic triangle within a milieu constructed by the teacher

Why, then, would anyone want to create a less certain, more complex teacher–student contract? The teaching in this book generally does not fit a simple version of the pattern just described, or if and when it does it is for particular purposes such as learning or relearning to use a notation, manipulation, classification, representation or a way to set out a calculation—precise use of the conventional tools of mathematics. These are normal things to do in classrooms and the teachers in this book tell students things when they need to.

In Brousseau's theory of teaching, the breaking of a typical contract (teacher shows and student follows) leads to new kinds of student engagement, not rushing to do some process that has been learnt and automatised but instead entering into the mathematical meaning of a situation (Brousseau 1997, p. 32). Brousseau's suggestion of how to break the contract is that the teacher should step out of the situation and educate students to think independently. The teachers in this book rarely step out, instead they engage firmly in the interactions, giving students the prompts, time and support for their thinking either individually or in whole class episodes. 'Wanting students to think hard' was mentioned by most of the teachers and sharing exploration of mathematics with students was common.

In John's classroom, the typical format of teacher first providing knowledge and follow-up practice for homework is turned on its head, but he does step out once discussion is taking place. Tom does not necessarily comment on the truth of students' mathematical statements—he often waits until the mathematical story unfolds further so they might disentangle themselves. Octavia recognises the shift students have to make in her lessons and so might give them a period of acclimatisation, gradually shifting towards them taking more responsibility for self-direction possibly via working with a neighbour to build confidence. Saul expects students to directly inform him about their efforts, giving the safety of 'eyes-closed voting' to do so. In return, students there trust them to take their viewpoints into account. Teachers find themselves repeating: 'I am really not interested in the answer; I am interested in how you worked it out' and then prove this interest by giving time to methods of working. Sharing and evaluating different methods often takes place in these classrooms.

In interactions with students, the teachers in this book break the normal classroom contract of expecting quick answers using learnt facts and methods that have been practised to fluency. Rushed thinking might lead to inappropriate actions based on incorrect assumptions and superficial similarities; intuitive responses can lead as easily to unhelpful as helpful reactions to mathematical problems (Fischbein 1987). Classroom behaviour learnt in the past might lead to habits of response that privilege speed over contemplation, yet the sources of many mathematical difficulties are not faulty conceptualisations but inappropriate, possibly rushed, decisions about situations.

Their reasons for breaking the habit of speed are not only about the nature of mathematics as a subject for discussion and the careful crafting understanding, but also because they see students as people in relation to mathematics, not only as imitators of curriculum mathematics but as active thinkers. Sandy and Trina both lack confidence in their understanding of the subject but have an instinctive approach to pedagogy that arises from caring about how children think, and what they think. That view of learning is common throughout Chaps. 6, 7 and 8. Most of the teachers talk explicitly about the need to be persistent because some students resist, or are uncomfortable or unfamiliar with, the expectation that students need to think hard about mathematics. The teachers know they are breaking a normal contract to which students have been acclimatised by their previous school mathematics experiences, or, for younger children, developing contracts in schooling that might be different from adult–child relationships before school.

At the start of the book, I asked whether 'upskilling' students in the kind of mathematics that is publicly understood to be important is more worthwhile than developing young minds so as to exercise their intellect. Does one exclude the other? Is it possible to do both? Care is not about softening the difficulties inherent in mathematics nor about reducing cognitive challenge, although at times it may be necessary to suspend the depth of the subject to compensate for what has gone wrong earlier in a student's life and learning, as Janet does. But mostly the teachers in this book manage both by focusing on habits of learning, habits of discovery and habits of inquiry even when the task is to work through a list of examples of the use of a technique. These processes are supported by interactions that are about reasoning, making connections, recognising what is relevant and how it is relevant. These are what I saw happening in lessons; these are what teachers talked to me about: care for the development of minds.

Because teachers are breaking a contract, removing earlier certainties, their care needs to be obvious to students so that trust can be built. Care is not an added extra only for students who are disadvantaged in some way; many of the situations I have written about are inclusive, teaching and caring for all. Tom, for example, cares about his high-achieving students by giving glimpses of mathematical horizons. Care for learning is seen as necessary for all individuals and is a holistic component of these teachers' professional lives.

Care for the Outside

Part-way through the book (Chaps. 5 and 6) I suggested that care for the community, care for mathematics and care for learning through relationships could be a useful combination. Returning to these after the explorations in Chaps. 7 and 8 has led me to expand the meaning of 'community' in two directions.

The first direction is towards the connections between school mathematics and authentic 'outside' activity. How do everyday experiences give purpose and insight into quantification, movement, position and shape? Of particular interest is a dual relationship between grafting mathematical abstractions onto 'outside' experiences and recognising abstractions in the structures of 'outside' experiences. Are trees used to practice using trigonometry, or do questions about trees trigger a need for trigonometry? The idea of entwinement (see Chap. 6) connects these different ways of seeing relationships between mathematics and material stuff, including stuff that is designed specifically to model mathematical relationships. In Honduras entwinement is embedded in a farming context; in Sarah's school entwinement is achieved in multi-disciplinary projects.

The second direction is to students' personal lives. Do expectations of home or community support school learning? How are mathematical ideas expressed in the home language? Can a student bring the whole of who they are into the mathematics classroom or are they expected to leave other interests, joys and sorrows at the door? Are they hungry? Are they anxious about their achievement levels? Do they need a place to feel safe and be cared for? In Lochville and ITS, individual learners are supported through yarning and multi-agency teams; several teachers invite students to hang out in their classrooms (e.g. Jasmina); Janet is sensitive to how trauma might be triggered; Octavia and Saul are sensitive to students' concerns about curriculum coverage.

Figure 9.2 is my way of grasping how care for aspects of learners' lives in a community relates to care for learning mathematics in educational spaces. While the teacher has control over some aspects of the classroom, the milieu is also shaped by the personal lives of learners and their prevailing experiences of mathematics outside school, so that the overlap of caring for learners and caring for mathematics—the shaded area—exists in parallel to outside influences and personal development of learners. Any idea of controlling cognition is an unattainable myth because of the whole life of the learner, but careful design and re-design of mathematical opportunities makes learning more possible and likely. There is the possibility of an inverse effect: a strong focus on care for the learning of mathematics can draw learners into a safe place, a distraction from outside life if one is needed.

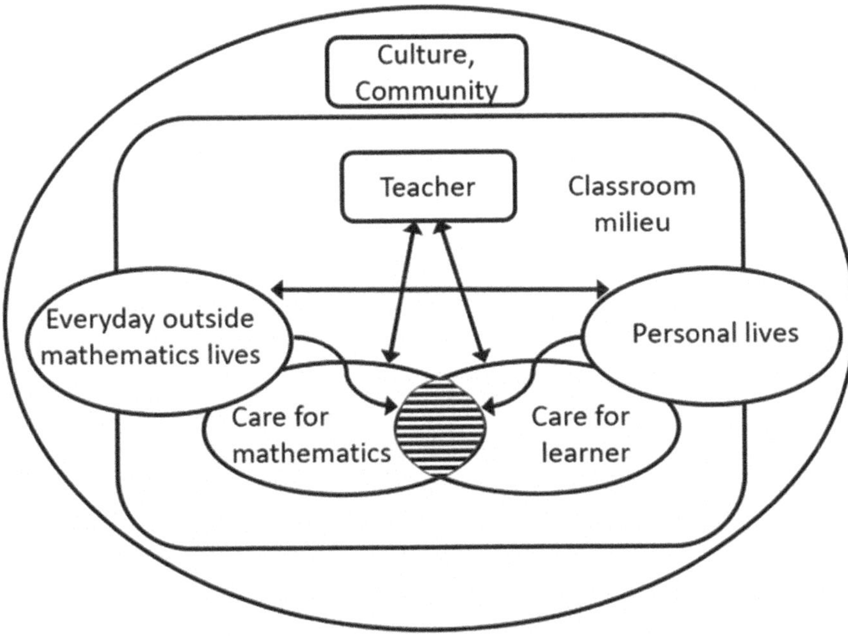

Fig. 9.2 An elaborated didactic triangle to show sources of influence on the teacher's care for the learning of mathematics

Trust, Listening and Relationships

Through listening we form mutual relationships with each other. Mutual respect is a feature of the relationships described in this book. Ultimately, teachers hope students will form sound relationships with mathematics through the pedagogic relationships. Florence's self-questioning moves from her own relationship with a mathematical idea to her students' possible relationships with the same idea, via her relationships with students.

In all of my examples, a major dynamic is that teachers listen to students' thinking and adapt their teaching accordingly, often working alongside students to achieve a mutual view of mathematical ideas. For the teachers in this book listening is not evaluative, not encouraging the student only to say something correct, but supporting students to say what they understand. There is no need to listen only for what is correct, because the teacher who knows the subject steers thinking towards a mathematical idea by well-chosen examples, representations and tasks.

Nicola's care is explicitly informed by understanding about how long-term memory is constructed from repeated experiences with multiple representations and access routes, and she has to contribute to students' formation of complex and useful concept images on which future ideas can be built. Access to these images has to be via fluency in language, facts and methods that trigger memory. If she does this well, they will trust her to guide their learning. All the teachers see that it is more valuable, and shows more care, to comprehend the students' perspective and work from there rather than to ignore differences and impose a new idea by coercion. Because this dynamic is often a contrast to previous school experiences, or to school or parental expectations, or to textbook design, mutual trust is the basis for a change in the teacher–student contract.

In return for trust (which works both ways) is autonomy (which also works both ways). All the teachers in this book exercise autonomy and also allow some autonomy of thought and action for their students. Autonomy for learners is about being free—even being encouraged—to do your own thinking and possibly to think differently.

Relationships in the Learning Space

The use of the physical environment can also influence students' relationship with mathematics and their teacher. Sarah talked about how timid children come to 'own the space' by hearing themselves speak about mathematics and knowing they have been heard. Around John's table students own the space while the teacher stays still and silent. Mark stands to one side so that students interact with each other and what is on the board, not through him. Tom squats down to student eye-level to learn about their mathematical thinking. If the Ojibwe word '*Gaamaamawiasigagindaasoyang*' were shorter I might suggest it be substituted for 'mathematics lesson' as it means 'gathering to do mathematics together'. Jennifer and ITS use circle time. In Janet's classroom, they have a place that is theirs and all their kit is there waiting for them. Students might gather round an expert of some kind, from outside, from another subject, or a teacher or peer who has something to show. Morgana and Jasmina open their classrooms at lunchtime as quasi-therapeutic spaces.

The environments used by Trina, Ed, Sandy and some of the educational spaces in Chap. 6 are not classrooms (Ed's includes the kitchen table) but natural activities, harnessed for seeing through mathematical eyes. The 'teacher' in any of these spaces might be alongside, inside or outside the mathematical activity physically or verbally. The word 'alongside' seems to have played a role in this book to offer a non-divisive, non-hierarchical way of seeing relationships between people in classrooms; between formal and informal mathematics; between the physical positions of teacher and student, or peer learners; between the need for directed learning and the development of inquiry and discovery habits; between academic, emotional and social development all happening alongside each other.

Care for Individual Voices and Thinking

Fundamentally, the teachers in the book trust that students are capable of the necessary thinking; this trust is in part dependent on their knowledge of how learning takes place and in part dependent on information from others, such as from research, that a particular pedagogic approach enables that kind of learning. For example, Nicola understands how to enrich long term memory; Mary understands the role of diagram; Saul uses variation to expose mathematical features. Many of the situations I describe depend on the fact that human brains have evolved to solve problems using a range of tools, to grasp and coordinate verbal, visual and physical information, and to adopt habits of seeking and expressing patterns and relationships among objects that they experience, including abstract mathematical objects presented in symbolic or diagrammatic forms. The use of a range of tools is seen clearly in Rathbridge, in Sarah's school, and many other cases. Spatial concepts used with young children are a particularly important manifestation of care for learning. Mathematical relationships, particularly those that relate to quantities and patterns, arise naturally from acting in the world. This was shown in Forest School, in the Honduras SAT project, in the Yup'ik project, and in the cross-disciplinary projects in Sarah's school. Teaching situations and talk nurture the development of logical and abstract thinking, and the emotions that support motivation.

Much of what I have said about care is to do with our sensory, emotional and social selves—the body, heart and spirit dimensions of being human. Care for learners' minds includes having adults who listen, comprehend what they say and how they think, and take them seriously. Adults who talk with them about mathematics offer not only the technical language and serial instructions for facts and methods, but also ways in which concept images might be constructed. Adults who listen can monitor how concept images are being constructed and adapt their own speech and future tasks accordingly. Language and listening are therefore major conduits of care for all aspects of students' development and particularly for mathematical thinking. Such thinking is emancipatory, either through gaining qualifications, or through becoming able to deal with the mathematics that arises in outside life, and also through feeling that mathematical thinking can be 'for them' as Mark says. Language and listening in the midst of mathematical opportunities are the mechanisms by which care for mathematics and care for learning are fused into care for the learning of mathematics.

Care for Mathematics

The teachers' view of mathematics includes the ability to see mathematical patterns and relationships in natural phenomena and in also physical and visual materials, and to use their knowledge creatively to solve complex realistic problems and move towards abstraction. This is most obviously seen in Chap. 6 in a mixture of authentic cultural activities (working with beaver pelts). In Octavia's articulation, mathematics is not about procedures, but about the behaviour of, and connections between, mathematical objects and therefore those should be the focus of her lessons. The teachers' view of learning includes respect for students' sense-making, and the belief that this can be harnessed throughout mathematics. This requires them to help students articulate mathematics. As Tom says, mathematical ideas grow like stories, and while the teacher has an overall sense of the plot, it is the students who must write the narrative. Cynthia Nicol, writing about work in Indigenous communities, describes coming to know anything as 'experiential, storied, relational, contextual' (Sterenberg et al. 2010). These apply as much to coming to know mathematics in classrooms as they do to everyday components of life.

Task design is not driven solely, or even mainly, by the logic of the mathematical concepts that have to be learnt, but by thinking through what students need to make a journey into those concepts and choosing or designing accordingly. Teachers select and design tasks that need discussion and mathematical reasoning as well as tasks that reveal mathematical patterns and structures.

Within these supportive, listening, educational spaces, the detailed attention to how mathematics is presented is often exceptional—beyond what a textbook can provide. The same text and task in the same font, visually and verbally the same, can generate entirely different interpretations from different students and give teachers insight into how students are building their concepts. Teachers respond by creating and adapting examples and representations as necessary. Thus, Saul gives students paper cut-outs of right-angled triangles to 'play' with before they need to recognise and create right-angled triangles in problems about bearings. He and Tom design their own example sequences and other tasks that draw attention to mathematical structures rather than procedures. Octavia gives exploratory tasks that she knows open the necessity for higher concepts. Ed poses questions, and helps his children raise questions, and then provides the precise mathematical tools that they need to answer them. One of Emily's students asked what month comes after December, and as a result the whole school changes from using linear models of clock and calendar time to using cyclic models so they can see that January comes after December. Sandy's tasks are mainly approximate because of the context, so she distinguishes between measuring, estimating and counting in task design and in her use of mugs. Teachers use examples that make access to mathematical concepts as likely as possible. Mark makes sure that students' thinking about circles is not slowed down by problems with substitution; Nicola talks explicitly about the importance of constructing long-term memories; others are less explicit but their use of images and actions shows appreciation of the need for mathematics to be meaningful. Even in Mary's fraction revision lesson images are used to give methods a diagrammatic form.

'Mathematical challenge' is a component of Jaworski's teaching triad (see Chap. 5). It is not a concept with a fixed definition but depends on identifying what mathematical challenges are evident in a teacher's work.

Failure, or feeling they are outside the flow of lessons, is a state that many people have in a mathematical environment, so doing any mathematics at all can be a challenge for some students. Janet meets this head on with two mathematical arenas, algebra and trigonometry, for which she structures experiential activities rather than announcing 'we are going to do X today'. She then gives social and emotional support, including sharing emotions with each other, coaxing them towards success. By demonstrating that they can do this work, and how it relates to exam questions, she begins to change their self-efficacy from failure to competence. Only when this has been achieved, over time, will she return to arithmetic where past failure has taken up residence and use anything they can remember as building blocks in a new atmosphere of success. In Sarah's work, mathematical challenge arises often in projects, such as 7-year-olds coping with scaling, when they have the will and need to know. In Octavia's and Tom's lessons, the tasks offer ideas beyond the current curriculum, and the immediate challenge is to move from conjecture to explanation and proof in a situation. In John's work, challenge is in the debates in class, and facing it is a shared endeavour between peers.

At the same time, there are fundamental understandings and operations that are needed, and the teacher has responsibility for ensuring these are learnt by all. Nicola will spend whole-class time publicly 're-wording, re-drawing, re-shaping, re-questioning' with one student if she believes this will be for the benefit of all, but also to show that it matters that everyone should 'get it'. Her stated philosophy is that her teaching should 'address the way they learn, rather than the other way round'.

Mary Pardoe in a private email to me, having read Nicola's philosophy, suggested another way to make this claim: 'teaching should be based on our understanding and knowledge of the way they learn, rather than dictate or force how they should learn'. For me, this touches on an essential difference of focus between cognitive control and cognitive care.

Professional Care

The teaching approaches described in this book address the conventional measures of success used in their institutions but for the teachers the overwhelming goal is mathematical confidence developed through learners' own thinking. The teachers are able to achieve this through personal commitment. Most have the support of their schools and colleagues. Some, but not all, of this support is through funding and smaller class sizes. Some of them manage with no extra funding and classes of 30.

All these teachers are engaged in professional networks, so are up to date with current ideas about mastery and cognition and they also maintain their own mathematical experience and learning. Task selection and design, bearing mathematics and learning in mind, orientate their lessons towards mathematical concepts and methods of inquiry, supported by confidence with procedures. The way teachers fill the area that integrates caring for learners and caring for mathematics is not through acting on efficiency and optimisation, yardsticks that are necessary when institutional constraints of funding, timetabling, curriculum coverage and testing regimes dominate pedagogy. Rather, they fill it with listening, talking, establishing trust. They provide deliberately planned contexts and they interact with spontaneous responses using tasks and interactions that value, encourage, respond to and expand learners' mathematical thinking. They sustain their work through pedagogical habits that are sustained by their belief in students' powers of mathematical thought.

References

Brousseau, G. (1997). *Theory of didactical situations in mathematics* (N. Balacheff, M. Cooper, R. Sutherland, & V. Warfield, Eds. & Trans.). Dordrecht: Kluwer.

Fischbein, H. (1987). *Intuition in science and mathematics: An educational approach*. Dordrecht: Reidel.

Jaworski, B. (1996). *Investigating mathematics teaching: A constructivist enquiry*. Abingdon: Taylor & Francis.

Sterenberg, G., Barrett, L., Blood, N., Glanfield, F., Borden, L., McDonnell, T., Nicol, C., & Weston, T. (2010). To become wise to the world around us: Multiple perspectives of relating indigenous knowledges and mathematics education. *Vinculum, 2*(1), 7–19.

Index of Teachers and Elders

A. Watson, *Care in Mathematics Education*, Palgrave Studies in Alternative Education,
https://doi.org/10.1007/978-3-030-64114-6

Index of Authors and Contents[1]

[1] Note: Page numbers followed by 'n' refer to notes.

A. Watson, *Care in Mathematics Education*, Palgrave Studies in Alternative Education,
https://doi.org/10.1007/978-3-030-64114-6

The manufacturer's authorised representative in the EU is Springer Nature Customer Service Centre GmbH, Europaplatz 3, 69115 Heidelberg, Germany. If you have any concerns regarding our products, please contact ProductSafety@springernature.com

Printed and bound by CPI Group (UK) Ltd, Croydon, CR0 4YY
22/09/2026
02221061-0001